Social Media and Social Movements

The Transformation of Communication Patterns

Edited by
Barış Çoban

LEXINGTON BOOKS
Lanham • Boulder • New York • London

Published by Lexington Books
An imprint of The Rowman & Littlefield Publishing Group, Inc.
4501 Forbes Boulevard, Suite 200, Lanham, Maryland 20706
www.rowman.com

Unit A, Whitacre Mews, 26-34 Stannary Street, London SE11 4AB

British Library Cataloguing in Publication Information Available

Library of Congress Control Number: 2015955902
ISBN: 978-1-4985-2930-3 (cloth : alk. paper)
ISBN: 978-1-4985-2932-7 (pbk. : alk. paper)
ISBN: 978-1-4985-2931-0 (electronic)

Contents

Social Media R/evolution

An Introduction

Barış Çoban

In contrast with other media revolutions, social media revolution is the one revolution which builds and shows itself over the streets. Although the evolution of social media has occurred in an intrasystem process, the revolutionary transformation in the use of social media has done so in an antisystemic way. The evolution of media proceeds in the context of developments in the field of informatics and technology, namely capitalistic relationships; therefore, revolutionary struggle goes on the basis of the creation of class conflicts and/or independent/autonomous social spheres. The struggle against the power in the field of media is being carried on by different means and methods. Social media revolution is one of the most important determinants of the struggle of carrying another world of today's beyond possible and the struggle of actualizing it. Accordingly, it's not possible to think of many social movements of the day without social media, or rather "alternative/activist new media" (Lievrouw 2011). Today's revolutions cannot be described only as social media revolutions: Traditional revolutionary forms of struggle, social movement methods and uses of traditional media are still utilized effectively, however, social media has become centralized in proclamation, organization, mobilization, and maintenance of revolutionary social activity by creating ample opportunities that other media tools have never held. Although the movements occuring in the local base had global impact starting with the Iranian Green Revolution (2009) and continuing with Arab Spring (2011), Spanish Indignados (Outraged) movement (2011), Occupy Wall Street (2011), Gezi Resistance (2013), afterwards, they organized, made themselves heard, mobilized, and gained recognition with an internationalist solidarity. These social media-based social movements built "temporary autonomous zones" on the basis of local occupy movements that included and transcended the antiglobalization movement and they established their own "dual powers which are not 'the powers.'" The movements that organize horizontally and create their own centers and participatory networks around this center built "hybrid-network society" (Amor et al. 2013, 4) or "hybrid public sphere" that consists of a mix of both the social media network and the network they created in their living space: "Egyptian activists are still trying to find a balance between the use of social media and local street campaigning. In so doing, they are creating a hybrid of the two, a mediated public space, as signalled by the appearance of Twitter hashtags as graffiti tags on the walls of Cairo" (Gerbaudo 2012, 74). Social media is a must for hybrid-network society in the formation process of a new political culture horizontally based on participation and pluralism. From micro to macro, social media plays a central role in the building and actualization of a new culture, politics, and lifestyle in a number of living spaces.

In the description of Castells, network society is a social structure that depends on the networks operating over information and communication technologies (2005, 7). When referring to the network society, social relationships operating in the basis of economy, politics, and culture are mentioned in fact: These relationships have a dynamic structure that consistently changes and transforms; however, the historical processes this structure has been through could also be mentioned. In this view "information age" makes a reference to a process in which economy, politics, culture, and all social thought and practices accordingly go through a reformation restructured over these networks. The power is also organized in the shape of a network and, trying to encapsulate the whole society, this network, works to keep the society connected to itself over communication networks. In opposition to the power networks' effort of establishing hegemony on the society, opponent movements create their own antinetworks and go into a struggle rendering the networks of the power noneffective or nonfunctional: "By engaging in the production of mass media messages, and by developing autonomous Networks of horizontal communication, citizens of the Information Age become able to invent new programs for their lives with their suffering, fears, dreams and hopes. They build their projects by sharing their experience. They subvert the practice of communication as usual by occuping the medium and creating the message" (Castells 2012, 9). The power of the opponent movements lies in their success in creating a network; the creation of a social network in the traditional sense was actualized in the basis of developing theirs against the strategies and tactics of the power and establishing their own communication networks against the networks of power in this process. It's also possible to read this process as a struggle between networks. Today, with the change of all network structures and new communication networks that have been created in parallel with this change, the struggle forms between networks had to transform. The antinetwork has become an international and effective power in contrast with the globalization of the network of the power. Rebuilding and use of antinetwork starting with the antiglobalism movement goes into a new phase with Arab Spring. In this process, antinetwork has become a more effective power than ever before. The creation of social movements with a rather effective organization and mobilization power has become possible with the emergence of the structure of traditional network and digital network in this process. As Negri and Hardt (2012, 90) state in different parts of the world, occupy movements put down roots on their individual conditions, took many things from each other's practice, and transformed them within the process as well; taking each other's slogans, but attributing new meanings to them, and most importantly, regarding themselves as part of a project. A powerful social network was created in the process in which the street actions link with each other in both local places and the international field over social media, with each practice interacting with the other, including and transcending it. The hybrid structure of this network does not exclude the street, although it locates the social media in the center. This new hybrid-network society became possible with the reflection of the network created over social media in the street. Social media activism turned out to be able to become an effective power when it reached out to the streets and squares. The fact that any opposition which remained close in the discursive field had no power to transform real life and that the actual transformation would be realized with actions was also recalled in this process: "The encampments and occupations of 2011 have rediscovered this truth of communication. Facebook, Twitter, the Internet, and other kinds of communications mechanisms are useful, but nothing can replace the being together of bodies and the corporeal communication that is the basis of collective political intelligence and action" (Hardt and Negri 2012, 21). While Castells defines Arab insurrections as the spontaneous processes of mobilization which occurred by the calls from Internet and wireless communication technologies by depending on both digital and face-to-face social networks preexisting in the society (2012, 102), he introduces the characteristic of this hybrid network. A movement that does not occur in the streets has no chance to be successful against the state. The street is the only medium where all the oppressed who cannot use the opportunities of social media can express themselves; for this reason, the mobilization of the struggle against power that starts on the basis of social media has to be maintained in different ways in the street. In this process, traditional forms of struggle and communication strategies are also used, and hybrid networks of power in which the opportunities of social media and traditional media cooperate are created. It was observed that social media and traditional media were hybridly used at different levels and forms in all of the street actions before and after Arab Spring (Gerbaudo 2012; Castells 2012).

The class structure of social movements which are organized over city-based and hybrid networks is another matter of debate. Today's social movements demonstrate not the fact that the class struggle is over, but that it has come back ascending in a new form. Transformation of classes and class trauma that the urban population has been through with workerization of the white-collar has also been active in this process; the masses which have been proleterianized, insecured, and made futureless show to both themselves and to the whole world that a new world is possible by experiencing. The new worker class looks for the ways to remove the enslaving oppression of the power in a different form from the past, works to pave a way for itself; a process in which the networks of power and alternative new media have a rather central role. Regardless of geographical location, with their common characteristics, diversifying resistance, organization, and communication practices since the distinctive conditions are being developed. These practices both include and try to transcend all the struggle forms of the past. In this sense, it is seen that the traditional political movements have difficulties in being related to today's social movements and that their classical articulation efforts fail. It could be argued that political transformation has become a necessity both for all the masses and all the political organizations in this sense. It can be argued that, there is a necessity for the masses to overcome the fear of the organization, and to see the truth that unorganized power will be occupied by the most powerful organized movement; to reform its Jacobin, monist, authoritarian, hierarchical, and non-participatory, nondemocratic structures in a suitable way for the process or to create new organizations and organization forms suitable for the process. In this process, creating alternative/radical media and social media and using them actively is of central importance, for the process of creating democratic, and participatory movement and creating a social movement in the same characteristics overlap, as Lenin (1987) mentions in "What is to be Done."

Arab Spring, indigenous movements, Occupy Wall Street, and other occupy movements and the Gezi Resistance are the processes in which the dual power states and autonomous zones are built, in these processes "the commune" that the activists built in the areas they kept for some time is a rehearsal of "dual power" (Lenin 1974; Bookchin 1992). In this respect, Colin Barker's (2008) "revolutionary rehearsals" focuses on the traditional social movements that have the aim to gain the political power; however, today's social movements can also be regarded as revolutionary rehearsals though in a different form with regard to creating dual power. Despite lasting for a short time, barricades, street wars, and communes suspend the power, and enable a new power in the street to blink. These suspensions, splits, fractions, and reappearances that can be described as "the blinks of revolution" probably refer to the short-time dual power instances. One of the main concerns of this process is the creation of network organization, that is, communicative network and social media were actively used in this process, in the process of creating the movement and using, maintaining, announcing, and expressing new kinds of horizontal and open organization forms, especially Twitter was actively used. In the resistances and insurrection in question, dual powers created their own alternative media and developed new communicative practices. The individuals and groups who were mobilized on the basis of "create your own media" or "be the media yourself" arguments made their voice heard both in real and virtual streets by using the opportunities of both traditional and new media. By experiencing the possibility of creating a movement that influences the whole country by a wide network over social media, the creation of a horizontal, participatory, and active network of media became one of the basic arguments of social movements. The media created in the streets, the new media, is reproduced over social media, and a hybrid media structure is built in the context of social struggle. For example, have become grafitti or vice versa. The transitivity between real and virtual streets radically transforms the structure of both social movements and media practices.

Today's social movements rebuild their repertory of actions on the basis of new media and social media thanks to the transformation in question, using new communication and action practices by hybridizing them with components taken from the past experiences. It could be asserted that successful practices were exhibited in Egypt and Tunisia in this process. Effective use of alternative media also carries importance with regards to the development of movements. Effective use of messenger services of Facebook, Twitter, YouTube, blogs, and mobile phones became irreplaceable for revolutionary organization and mobilization in the process called "social media revolution." From now on, activists use Facebook in order to plan the actions, Twitter to coordinate them, and YouTube to announce them to the world in all of the actions today (Howard 2011). Social media provides new revolutionary actors with opportunities necessary for them to perform their actions: it first creates a space where the activists identify and express themselves, then it enables these activists to share their ideas and productions and helps disconnected activists share decision-making and coordinate, consequently making them come together and mobilize. Social media reduces the time and cost of coordination for activists and enable the circulation of information and literature. Social media produces a common sense of awareness, even create an "imaginary community" over this coordination and circulation. The creation of this imaginary community was possible thanks to the creation of a sentimental and intellectual commonality between the activist in the street and the sympathizer-audience online. Social media aided in the creation of a network connecting inside and outside, mobilized the anti-publics who supported the occupation areas both materially and spiritually: "Social media only acquired importance during the phase of sustainment of the movement, being used to create a sense of attraction to the occupations, and to invoke a sense of solidarity between 'physical occupiers' and 'internet occupiers', activists on the ground and people following events from a distance" (Gerbaudo 2012, 103). In this respect, providing a massive support that had not been possible in the past was enabled by effectively communicating with activists and non-active audience-sympathizers apart from the communication and interaction between activists.

Although Arab revolutions are called social media revolutions, they are actually the ones that used social media effectively, and which are an extension of revolutionary surges coming from the past. Transformation of communication forms correspondingly leads to the transformation of actions, for social movements the same are the communicative actions. All of today's social movements can express themselves via alternative new media. The "Green Revolution" starting in Iran in 2009 demonstrated via Twitter that social media was also quite significant for the social movements in the Middle East, yet it was irreplaceable due to the control and oppression of the powers. Media of authoritarian regimes, or the influence of powers on mainstream media, their homophonic supporters are vitiated by the effective use of new media. It could be argued that there is no period in history that the mainstream media has been through such a trivialization and a fall from grace. Alternative new media enabled resisters to create their media and the creation of a democratic, participatory, and collective structure of media that grants the activists the status of media producers. The masses who rescued Hugo Chavez, the president chosen by the prevented coup d'état supported by the United States in Venezuela in 2002, from subversives put forward their rebuttal against the mainstream media, saying "The revolution is not going to be broadcasted on TV." Improving their new alternative media against the mainstream media, the opponents revolutionize the field of media as well. "The revolution will be tweeted" has been written on the walls of the rebellious cities from then on; resisters produce their media and channels in the streets, in resistances. Media thus transforms into an integral and indispensable part of the activists' own practices.

Media is a sine qua non in the development process of anti-systemic movements, for the massification and mobilization of the movement are possible with the media. Social media functions as alternative media on the part of the opponents. Activists use social media in order to share the photos, videos as a form of alternative media with the exploitations and the narratives related to the corruption of the government (Barrons 2012, 57). Presenting the crimes that the powers committed against the masses, these publications bring it into view that the struggle against the power is not much possible without alternative new media any longer. Alternative new media, exposing and archiving all the crimes committed by the power, is the new channel of social memory at the same time. In spite of all of the censoring, prohibiting, and restraining efforts of the power, digital activists achieve overcoming the obstacles by building barricades in the streets of the Internet, moving forward underground. In this respect, the struggle of activists continues in both real and virtual streets.

The structure and operation of social movements have undergone a transformation with alternative new media, especially with social media. A polycentric, horizontally networked, flexible, and dynamic structure has replaced the centralized, hierarchical, strictly ruled structure in the past. Since the opportunities that social media have allowed for the social movements to be created and maintained with a participatory, democratic approach, social media has enabled a great number of people, who have stayed outside the political sphere so far, to be politicized and to get involved in the practices in which they can express themselves in this sphere. Social media caused a new period to begin in the process of development of the new forms of politics and political struggles in this context. The powers' struggle with social media is one of the most significant indicators in this sense; the efforts of the oppressive powers to establish control over social media and to restrain it arise from the fact that social media has made it visible that it is possible and even compulsory for social media to create new, effective, meaning having the power to transform the daily life, forms of politics, and social freedom spheres against the power. The necessity of the state to build up counteractions against social movements becomes evident; the police and intelligence organizations are reformed, new legal regulations are prepared against opponents, and new prohibitive legal regulations about the field of new media are arranged. Anti-revolutionary organization of the state is nothing new; it is the organization of a dynamic, offensive instrument of oppression, favorable for new offensive policies of neoliberalism and having similar forms almost all over the world. A police organization, which has been specialized in responding to social events and has new weapons and equipment, intensively uses "aestheticized violence" (riot control vehicles, panzers, gas bombs, robocop clothing etc.) with the support of mainstream media. Police organizations develop their own counteroffensive forms by analyzing the social movement practices in different countries. In return, activists share their practices of new media and build up self-defense forms against state violence. These self-defense forms involve civil resistance and could involve "limited counterviolence" as well. In this context, it should be noted that violence is a form of communication. It is possible to act on the basis of passive resistance in order to destroy the violence monopoly of the state; moreover, the use of legitimate violence may be compulsory case by case in terms of being effective in the social sphere and constructing free/autonomous zones. “Limited counterviolence” can be described as the use of violence on grounds of self-defense against the violence of the system without giving any harm to a government official or giving as less harm as possible. Opponents determine the levels and forms of violence, according to preexisting conditions with respect to the violence level of state. Besides, the actions which demonstrate the desperation of the state before the society by sacrificing itself instead of responding to the violence of state could be mentioned. The young activist who burnt himself to death in Tunisia and the elderly opponent who committed suicide in Greece took a stand against and barricaded the state violence turning to the social body by confining the violence to their bodies. It could be claimed that this type of action has an undeniable potency and it is the modern reflection of Fedayeen culture. Tunisian revolution gained momentum with the “violence” of the sacrificing action Mohammad Bouazizi performed by burning himself; the wall of fear was overcome and the path of massive resistance was cleared. Active use of social media has a major role in the action of Bouazizi to become the starting point of the revolutionary grassroots movement in this process as well. Similarly, the Facebook page created for Halid Said who was killed as a result of police violence in Egypt was one of the determinants at the center at the beginning of revolutionary resistance. The young ones killed by the police during the Gezi Resistance in Turkey have become the names and faces of the new revolutionary spirit in both real and virtual streets.

Legitimate resistance against violence, active use of the forms of action and resistance which is non-violent and reveals the violence of the state in this sense, enables social support to be massificated. It is possible for nonviolent and civil resistance to be effective when they have gained. Making visible the civil resistance by using the alternative new media actively and the “disproportionate violence” that the state uses against it, undermines the legitimacy of the state and force the state be in solidarity with the resisters who are subjected to violence, causing public indignation. Social media has become a substantial medium in revealing the state violence and making the passive resistance effective; as explicitly demonstrated in the Egyptian and Tunisian practices. It could be observed that the active use of social media can draw and mobilize social support as well as it can downgrade and disable the police violence besides revealing this violence in many cases. Consequently, social media becomes an effective power in permeation of non-violent social movements.

The Gezi Resistance refers to both a new revolutionary process and the social media revolution. Following the footsteps of Arab Spring and subsequent occupy movements, the Gezi Resistance turned the political life upside down, and its influences have still been going on. It can be argued that with the Gezi Resistance, the social opposition in Turkey started to undergo a great transformation, and it has involved all the "occupy" movements and revolutionary spirit of this transformation, going even beyond in some respects. The Gezi Resistance has a structure and operation beyond the usual political movement repertory in Turkey; thus, traditional revolutionary opponent circles failed to catch up with and lead to this movement. This new social movement in which the rules and habits were broken demonstrated that it is possible for a young population which have always been apolitical and passive to be politicized and begin to resist. Furthermore, "Gezi Commune" showed that the all other occupy movements could be kept alive accordingly by establishing the communal living space in an area to be regarded as the heart of the country. Commune practice is of capital importance, for it keeps the clues that a new independent country can be shaped beyond the rebuilding of the city. In contrast with the approach of traditional revolutionary movement that ignores the attempts to transform the life now and postpones the transformation until after revolution, it was shown in the practice that the claim to implement the revolution in all spaces of life. The creation of temporary autonomous zones (Hakim Bey 1991) or "dual power" (Lenin 1974; Bookchin 1992) spheres and the competence of keeping a communist alternative living space alive for two weeks in the city center that can be considered the heart of the country demonstrated that revolutionary dreams could be realized in different forms than today's. Defended by the barricades, Taksim Commune illustrated that the revolutionary option had not come to an end, that the traditional revolutionary practices had not become functionless, and that the new struggles would arise upon the traditional ones. The way for the development of new revolutionary discourse and practices was cleared with this process. In this sense, the possibility of the invention of a revolutionary and leftist option that is horizontally organized with the participation of the masses is realized. It is also seen that the creation and activation of alternative new media are of central importance in order to achieve it. Alternative new media has become an active power in a way by which it interacts with traditional media and also plays a significant role in developing new practices of struggle.

The Gezi Resistance is a "civil movement" like both Arab Spring and occupy movements. It emerged in the basis of "speak" against the oppression experienced in both bringing itself into existence before the power in general and expressing itself by occupying areas. Being the spontaneous movement of the masses in a process in which the traditional socialist movement was having a crisis in reforming and going beyond itself, the Gezi Resistance built itself on the basis of social media. Basically depending on white-collar workers, the students and the unemployed as the civil movement, this movement operated by the active use of social media, and especially Twitter. As Scholl (2012) mentioned, the Gezi Resistance is utilized directly as civil disobedience and simultaneous action as in the case of the antiglobalization movement. The masses kept going in a line that rarely exceeded the limit of legitimate action despite the intensity of the police violence. In Turkey, the Gezi Resistance made a different progress from traditional revolutionary mass movements and led to the start of a process affecting and transforming the whole society in which unusual new forms and tactics of the struggle were used. The confidence in mainstream media was completely lost with Gezi, and all news was followed via alternative media. Social media, Facebook, and Twitter can be said to have been intensely used both for communication and transmission of news. Social media was used to organize, to perform (Twitter was an effective medium in the process of playing puss-in-the-corner with the police and that of construction or defense of barricades) and to establish intercommunication of actions at the same time; in this sense, a social media revolution was really experienced in real terms in Turkey. Spreading fast in Turkey, the movement made progress toward an antisystemic line besides being in a line against the power. A process was experienced, in which the fact that the state oppressed the Kurdish people and the other minorities for years became visible, and the dark side of the state was made visible. In terms of revolutionary struggle, this process can be read as "a tiger leap"; a process in which the foundations of the social structure that protected by the state with oppression were shaken, demonstrating that the potentials of alternative power can be implemented. In this sense, the state of dual power is of capital importance, although it was experienced for a very short time and in a weak way in Taksim. It became assertable as a revolutionary claim that the communal order the people created by relying on their power in a very short time without any organization could be implemented across the country. Being the practiced example of Communist theories which have been claimed to become invalid, the Gezi Commune can be one of the most explicit statements of the fact that "the Communist idea," which is the name of "the ability of an ultimate goal, not an ideal of society, a tangible and proceeding politics to conceive itself as a thing that aims the local, the partial and the universal" (Badiou 2012, 12), is still prevalent.

In conclusion, the metamorhopsis of the revolution starting with Arab Spring began a transformation process which influences all the communities in the region, and which is irreversible. The revolution, for which there had been no hope until then, was out in the streets in the geographies where it was possible. From now on, the necessity to mention again the alternative geographies and maps is obvious. Experienced practices have presented that the forgotten geographies of the world cannot be ignored anymore in the context of international revolutionary struggle. Experiences indicate that the social struggle that can break down capitalism could reappear in different forms in different geographies, yet it could agitate the whole world. The revolution uncovers the necessity of a collectivist social culture by consistently changing, transforming, and undermining the local powers. Alternative new media can be claimed to play an important role in this process as well. Social struggles that are carried out on alternative new media networks and in the international context can reach a level that could affect the powers. The alternative media has a more significant role and function than ever before; especially the use of the opportunities of new media and social media brings thus about a radical transformation in the context of both theory and practice.

The practices in which dual power states are experienced and social movements sustain their own powers in the basis of autonomous zones, albeit short-lived are built on the basis of communicative networks, and all the practices are shared in interaction with different locals as well. The occupy movements, in which Communist practices were experienced, and the rural and civil movements, indicate that a utopia that is expressed by a wandering ghost across the world is realizable. Beyond speaking, activists enable the sustainable revolutionary state and the uninterrupted social, political, and cultural transformation to be experienced with the help of the struggle forms are based on transforming their living spaces and implementing communist culture. Political goals are directly implemented now and are not postponed to the future, unlike in the past; the revolutionary movements rediscover daily living spaces and carry out the struggle of creating alternative autonomous zones. The social movements performing the struggle of creating "rebellious cities" in both real and virtual worlds, and actively utilizing the alternative new media, are becoming an active power across the world. In this process, social media, once seen as a channel of "word play" by the powers, has been made the "playmaker" of a revolutionary carnival. The social media revolution has brought with it the fact that the creativity of the oppressed can overcome all the obstacles be they heard or visible.

A ghost wanders upon the world—the ghost of social media. This ghost speaks the language of a new world and from now on, this language is spoken in all of the streets, from virtual to real. The information and communication technologies neoliberal capitalism has used in order to cross all the boundaries existing in the world have been transformed into the instrument of new revolutionary process in the hands of the masses that simultaneously struggle against it. Marxist, anarchist manifestos are reread, and the age of insurrections/revolutions aiming to radically transform the life from now in order to create another world is beginning. Slogans are chanted shoulder to shoulder in barricades, paving stones are removed in order to counter attacks, communes are set up in city centers. The same revolutionary practices are repeated in digital streets as well. Facebook, Twitter, YouTube, and the other social media networks interconnect all streets of the world, and the revolutionary wave spreads over these networks.

Social media revolution has started to turn our world upside down by influencing all living spaces. Many insurrection/revolution movements, including Iranian Green Revolution and Arab Spring following it, Occupy Wall Street Movement, Indignados (15M) Movement in Spain, and the Gezi Resistance in Turkey are called "social media revolution." With the social media revolution alternative maps showed up; occupy movements in Morocco and Nigeria denotes that a new insurrection culture has originated in many geographies the world has not seen to date. Although social media has the emphasis on individuality, it has enabled a social mobilization with the network it created, which has not been possible untill now. Therefore, it transformed political life in an irreversible way. Social media paved the way for political participation and action. Moreover, social media enabled the information the powers endeavor to make invisible to be revealed to world public opinion by "hacking" or "leaking" it. It introduced itself as a power having the ability to make a lot of secrets visible that the powers endeavor to hide and accelerated the development of digital activism on behalf of the society as a social supervisor having international support and legitimacy. Social media-based digital activism has central importance to perform social movements in different levels and forms; for example, factory occupations; defense of squatted houses; struggle for lesbian, gay, bisexual, transgender, and intersex (LGBTI) rights; environmental movements; movement of animal rights (e.g., Wild Horse Defenders), and so on.

Social media provides significant opportunities in the development and realization of radical political thought in both social and individual means. The secret of enemization of social media by the powers lies in the power of these opportunities. Social media activism has an important role in the development of democracy in the public sphere; in addition, it provides for the issue of rights and freedoms to remain on the agenda and aids in the effective realization of political participation—especially the political participation of the young—as well. Social media-based political campaigns force political systems to make democratic reforms in context of social demands by the support of public opinion.

Social media influenced working hours and caused spare time to transform radically, as well. Digital culture brought forth the reformation of a number of types of social relationships. The panoptical culture of social media that also normalizes being monitored and monitoring caused many problems related to privacy to arise in this sense.

Social media also radically transformed the educational sphere and learning practices. It can be remarked that alternative practices of education and learning have started to gain importance with this transformation that breaks the rigid structure of education. A society without schools is not a dream any more; our whole life transformed into alternative schools over the social networks. Social media provides opportunities for everyone to produce and share information without the intervention of the power in the context of life practices—although the poor do not have the chance to use these opportunities due to the digital gap created by class inequalities.

Social media, which is today's haunt of media evolution, is the crucial point of a revolutionary leap. Nothing has passed away in this "end of the world" in which evolution and revolution interwined; instead, everything has just started in that "this is only the beginning."

REFERENCES

Amor, J. D., J. Foss, G. Gkotsis, H. Grainger-Clemson, E. Marchis, and F. Azhar. 2013. "Modelling Social Mobilisation—An Interdisciplinary Exploration of Twitter as a Mediating Tool for Social Acts and Information Networks." *The Warwick Research Journal* 1(1).

Badiou, A. 2012. "İdeası Komünizm Olan Şey." *Komünizm Fikri,* ed. S. Zizek and A. Badiou. İstanbul: Metis.

Barker, C. 2008. *Revolutionary Rehearsals*. Chicago: Haymarket Books.

Barrons, G. 2012. 'Suleiman: Mubarak Decided to Step Down #egypt #jan25 OH MY GOD': Examining the Use of Social Media in the 2011 Egyptian revolution." *Contemporary Arab Affairs* 5(1), 54–67.

Bookchin, M. 1992. "Libertarian Municipalism: An Overview." *Society & Nature* 1.

Castells, M. 2005. "The Network Society: From Knowledge to Policy." *Network Society,* ed. M. Castells, and G. Cardoso. Washington: Johns Hopkins Center for Transatlantic Relations.

Castells, M. 2012. *Networks of Outrage and Hope: Social Movements in the Internet Age*. Cambridge: Polity.

Gerbaudo, P. 2012. *Tweets and Streets: Social Media and Contemporary Activism.* London: Pluto Press.

Hakim, B. 1991. *TAZ: The Temporary Autonomous Zone*. New York: Autonomedia.

Hardt, M., and A. Negri. 2012. *Decleration.* New York: Argo-Navis.

Howard, P. N. 2011. "The Arab Uprising's Cascading Effects." *PSMag Mazagine* http://www.psmag.com/politics/the-cascading-effects-of-the-arab-spring-28575/ .

Lenin, V. I. 1974. *Collected Works 24.* Moscow: Progress Publishers.

Lenin, V. I. 1987. *What Is to Be Done.* New York: Dover Publications

Lievrouw, L. A. 2011. *Alternative and Activist New Media.* Cambridge: Polity Press.

Scholl, C. 2012. *Two Sides of a Barricade.* New York: SUNY.

Shirky, C. 2011. "The Political Power of Social Media." *Foreign Affairs* 90(1), 28–41.

Chapter 1

Commune, the Web, and the Anarchist Thought of Mikhail Bakunin

Chiara Livia Bernardi

On the occasion of the insurrection of the Paris Commune in 1871, Mikhail Bakunin wrote that a revolution "could neither be made nor brought to its full development except by the spontaneous and continued action of the masses, or the groups and associations of the people" (Dolgoff 1972). In his ideal of a society free from the strains of the state, Bakunin emphasized the crucial element of free association, liberty, and respect of human beings and called, rather than only hoped or theorized, for a revolution of the masses—a revolution that would bring the destruction of the system and any form of institutions and control. Despite his striving activity of proposing alternative forms of societies, the work of Mikhail Bakunin is rarely mentioned nor used to analyze the current debate of the relationship between technology and society. This debate primarily moves along an obsolete track of paranoia versus enthusiasm or fear versus fanfares. Wendy Chun has investigated the process that brought the advent of the Internet from a very niche tool (initially used to connect the academia and the military) to a new virtual Far West, a place open to conquer to any free man willing to rush into an unknown territory (Chun 2006).

The Internet, and subsequently the World Wide Web and Web 2.0 have the common denominator of having been welcomed as the realization of a cyberspace or that of a genuine Big Brother in Orwellian terms. More specifically, in the optimistic sense this achievement of a cyberspace is seen as a far reaching but at the same time liberating "something" that transcends gender, race, age, and any other biological, patriarchal, or social constraints. In these dichotomies resides the debate of power, in its many different forms, whose essence now resides in the ephemeral yet fully tangible reality—of the code. The new hidden character of power, Chun argues, has given way to applications of the panopticon of Foucaldian memory and is strictly linked to control, passively exercised by the paranoia of knowing that any information can be easily controlled, processed, filtered, or even blocked (Foucault 1995).

However, Chun argues that this view has been opposed and almost contradicted by what could be defined as "Web Optimism," made up of an enthusiastic evaluation of the Internet, its democratizing potential and global impact in terms of spreading of knowledge and the bettering of people's lives on a worldwide scale.

These dichotomous, although relevant and theoretically valid, positions, have started to be shaken by a more critical analysis of what constitutes the relationship between technology and society. Such an analysis has shown that this relationship is better represented as a self-refilling circle: a network, or a rhizome, where nodes are continuously engaged in processes of negotiation, resistance, and agreement. French philosopher Bruno Latour has defined such a networked relationship as a sociotechnical imbroglio (Latour 1999). Scholar John Law instead favours a topological analysis, or a "local" approach to any subject of interest, that reflects on the complete enmeshment of society and technology, overriding any dualism effectively moving toward a more comprehensive and realistic approach to the current reality (Law 1986).

Finally, when embracing the concept of the full enmeshment of technology and society it is necessary to realistically recognize how they are imbued in each other. As a result, it becomes possible to propose a view of certain situations as being (and existing) solely through the merger of the technical and the social, with the latter having the ability to act as a political stage of action. What this chapter aims to achieve is to move this approach a bit further, re-reading the concept of association as proposed by Bakunin, and merging it with the involvement of the hacktivist groups Telecomix and Anonymous in the Middle East, focusing especially on the Egyptian Revolution, deemed by many a social media revolution or a Twitter revolution (Shirky 2011). It is in this aspect that this chapter will propose the unusual philosophical marriage between John Dewey's formulation of the term "issue" and the concepts of "commune" and "voluntary army" as proposed by Michael Bakunin, in an attempt to offer an alternative view of the events of the Egyptian Revolution in 2011.

JOHN DEWEY AND MIKHAIL BAKUNIN

John Dewey championed the idea of the issue as a very peculiar and local problem that materializes through the association of people affected by it. Once an issue arises, the goal of the gatherers is to solve it by addressing it to experts or anybody deemed capable of dealing with it, or the addressees. The primary characteristic of Dewey's issue is its flexibility: the issue arises and comes to life when a certain status quo is altered and therefore equilibrium is broken. Whoever is involved and harmed by the rising of the issue rushes into it and tries to find peers to discuss it, peers that share the same sort of affection and effective or potential damage. As a consequence, a public is formed. However, such a public is not "fixed." Its members can join and then go away at any time; each member can be part of other groups and have other interests; the public only exists as long as the issue exists. The common denominator behind the existence of a certain public is the issue, or the problem that needs to be solved. Once the issue is solved, the public disassembles, and each member that constituted it moves away, keeping its fluid characteristics of being one and many at the same time. Another characteristic of the issue is that it is a priori non-political. For the purpose of this analysis, politics will be defined in a classical sense, or as the "practice of governing a country local politics" and / or "national politics [...] or of governing a country" (Collins 2004).

Contrary to Dewey or any other theorists, Bakunin never really worked towards the definition of a coherent theory. His theoretical positions are presented rather loosely in numerous letters, pamphlets, and speeches that have to date received little attention in academia. As Brian Morris has argued, "the work of Bakunin was not one of systematic theorisation of ways to reach liberty for he was too much involved in revolutionary activity to have the time, like Marx, to sit around and write systematic works" (Morris 1993). Much of Bakunin's influence has come from his letters, which almost always had the length and the form of pamphlets. His literary output was enormous, and yet there is not one single completed and finished book. Nevertheless, "the importance of Bakunin is that he put no effort in being systematic" (Morris 1993). Despite the fragmentation of his theoretical corpus, Bakunin is a very prolific and coherent thinker, one that promoted, before Elisee Reclus and Peter Kropotkin, an ecological approach to the world stressing the never-ending link between nature and society, and the individualist and social nature of the man (Lehning 1973; Maximoff 1953).

In Bakunin's view, the individual is a creative agent, one that is defined by and defines the social and natural world; the world itself is seen as a creative process, one that includes the enmeshing of natural laws and social milieu. In consequence, it is in the recognition of such a creative process that Bakunin posits his concept of liberty as a non-absolute principle, different from free-will in its very essential nature. Liberty, Bakunin argues, depends on many different elements and variables. It depends on the man's obedience to "natural laws," not because they have been externally applied, or "imported" to use Bakunin's terminology, but because they have been recognized as necessary and good. Liberty is a purely social process, one of interaction between men in the name of mutual respect and good governance. The self-recognition and acceptance of natural laws is perhaps more revolutionary than any other theory of the time, because it not only denied the existence of a *Deus ex Machina* of divine nature but also denied any collective deity created in the name of ideology (Lehning 1973). Therefore, liberty is the outcome of society, it is a process that sees men coming together and developing their intrinsically social nature because man in isolation can have no awareness of his liberty. As a result, liberty is a feature not of isolation, nor of exclusion, but of manly interaction and connection. As Bakunin wrote, "I am properly free when all the men and women about me are equally free, far from being a limitation or a denial of my liberty, the liberty of another is its necessary condition and confirmation" (Lehning 1973).

Bakunin was a holistic thinker with local focus. Contrary to many other thinkers, he constantly highlighted how any revolution could become a social revolution only and exclusively if its local nature and causes would merge with universal ideals of liberty, freedom, and equality. He acknowledged and defended the protests of the Slavs, of the French Commune. He fought in Bologna in 1874 and supported Errico Malatesta and the small Benevento Insurrection in 1877. Bakunin refused affiliations and all-encompassing theories and ideologies, which he deemed as dangerous as religious creed. In fact, although respectful of Marx's ideologies, he refused to be called a communist. Similarly, despite his support for Garibaldi, he was also intimate friends with Errico Malatesta, who exalted brigandage and violence as means of actions against the new social order established after the unification of Italy in 1860. Bakunin never laid out a theory of technology and, contrary to Marx, never really embraced the concept of production as the cause of all evil or science as an imposition to control nature. He recognised capitalism as a concept that would throw off center the already fragile equilibrium of society but never actually theorized around it. The lack of theorization and systematic problematization of social, economic, and political issues make Bakunin's thought very malleable and flexible, capable of generating thoughts with regard to the current world of affairs without falling into dichotomies or conceptual entrapments.

Having seen Dewey's proposition of the issue and Bakunin's recognition of the local and peculiar nature of a social revolution, it is possible to say that an original analysis of digital cases of activism, online gatherings, or active and fluid publics that gather around issues would be best addressed using a combination of these two analytical views. Dewey's *publics* merge perfectly with Bakunin's idea of voluntary associations, or federations founded on shared ideas, principles, and beliefs. The same way that Bakunin never really defined how a voluntary federation would come to life, Dewey never defined how publics could take action. In both cases, the lack of completeness offers fertile ground for an analysis of today's issues, associations, movements, and even hacktivism. In fact, the spread of new technologies has allowed for new forms of aggregation, discussion, and disobedience to materialize. However, these should not just be interpreted as a transposition of offline publics into their digital alter ego. Different forms of associations, as offered online through technologies and antitechnologies of resistance, give new life to the Deweyan conceptualization of the issue and to Bakunin's ideas of voluntary federations, based on shared values and principles, whatever their nature (Brener and Schurz 2009). Finally, as it can be seen, an entire spectrum ranging from groups and forums, Web 1.0 to Web 2.0 platforms, including hacked websites have been fully involved in the process of group formation and issue articulation.

When moving towards the recognition of the digital transformation of today's world and issues, Bakunin's concepts of local causes and universal principles become fruitful sources of theoretical inspiration. Our increasing dependency on technologies and the transposition of information and communication into a set of digitized data should leave questions concerning the effects of technology over society and vice versa. Rather, what should be highlighted and investigated is the emergence of "natively digital" realities and issues. Addressing an issue or reality as "natively digital" is not purely related to the fact that it can be found online, lives online, and can be archived online. A natively digital reality is here intended as existing through, in, and thanks to the full enmeshment of broader structural apparatuses, made of social, political, and economic complex apparatuses, without neglecting the infrastructural element, which is instead made of fiber optics and cables, routers and Internet exchange points; databases, algorithms, and code. This rather rich and diverse plethora of actors engages in continuous negotiations with regards to power, control, rejection, and resistance, no longer solely exercised through geographical, institutional, and other "traditional" means but also through the complex and multiple layers that combine and constitute the digital.

Therefore, the very idea of the state as a set of institutions dictating the rules for a geographically bounded people fades away; its institutional power is questioned by a hidden element of power—the code. The code is here intended in its hidden although tangible characteristic. Code responds to algorithmic laws and, although human readable and programmable, it still pertains to the world of the machine, trying to express its full performativity in a behind-the-scene and user-unfriendly fashion. This essential requirement makes it almost impossible to understand where the human agency is placed and, rather, it sheds a new light over the discourse of agency, which is no longer solely placed with the human subject but it also includes the artifact; in fact, the code allows any technological artifact to make sense, to work and express its fully visible and user-friendly performativity. Such agency delineates the tangibility of code and its performativity, or the complex set of tactics put in place to activate, alter, change, stop, or allow something to happen.

In the specific case of the Egyptian Revolution, it would not be correct to state, as mainstream media put it, that the Internet was "blacked out" or "shut down." Instead, what made the Egyptian digital world stop on January 27, 2011 was the border gateway protocol (BGP), or more specifically the set of codes that allows the BGP to grant permission to certain paths to be passed through or filtered. This would therefore mean that the traffic coming from anywhere outside Egypt was rendered completely unrecognizable and passed as irrelevant. Any reflection about the revolution should therefore shift from *who* activated the block to *how* the block was allowed, resisted, and to some extent neutralized through various tactics that brought to a partial restore of the Internet in about five days. Therefore, when the natively digital nature of current affairs is recognized, and there is no pressure to "fit it" in structured theories or inflexible frameworks, Bakunin's reasoning come to support an analysis of current phenomena of online dissent, hacktivism, and political unrest. As with many other theories and approaches, Bakunin's thought was inflated by the romantic rhetoric of revolution and social change that spread in Europe at the dawn of the nineteenth century. In part, this exaggeration was the result of the disappointments that followed the French Revolution. Nevertheless, this and other aspects will be taken into consideration, and some of Bakunin's stronger claims will be analyzed from a more modern perspective and interpretation.

In Bakunin's thought the state as an apparatus of national sovereignty ought to be dismantled and replaced by federations of communes whose leitmotif would be free association in the name of the ideals of freedom and equality. This reasoning never aimed to purposely destroy social, economic, and political apparatuses in the name of an alternative (i.e., which could be the soviet, the proletariat or the capital). Instead, his ideal was that of a complete wipeout of the current status quo through a social revolution, a revolution that would have taken place only and exclusively when people (the masses) would have decided to take control of their own lives and futures, and to fight against any form of oppression. The state is intended negatively as the corpus of external laws, decisions, and constraints forced upon the people (or the masses) and not recognized as useful by them. In consequence, Bakunin's thought has often been mistakenly interpreted as violent and as aiming to create chaos, destruction, nihilism, and brigandage. Although it is not the intention of this paper to redeem Bakunin's thought from the systematic work of demagogic ridicule, this analysis will revisit some of these categorizations when examining the actions of the hacktivist groups Anonymous in Tunisia and particularly Telecomix in Egypt.

HACTIVISM: RESHUFFLING BOUNDARIES FROM HARDWARE TO SOFTWARE

In light of several recent events that have involved hacking, controversy, and unrest as their central themes, it is necessary to define hacktivism while addressing its connotation stemming from the combination of *hacking* and *activism*. Tim Jordan and Paul Taylor defined hacktivism as "activism gone electronic," immerged in the cybertext, and speaking the cybertext (Critical Art Ensemble 1994). Hacktivism is therefore a combination of political action and coding knowledge. In 1993, the Critical Art Ensemble (CAE) developed a project called *The Electronic Disturbance* that defined hacktivism in terms of geography and spatiality, maintaining that the hidden character of power could be finally openly resisted through and within the fabric of cyberspace. The term has recently been associated with rebellion against the superimposed order of governments, regimes, corporations, and financial institutions, and whose main weapon of mass destruction is the code.

Hacking is not a recent term and, although it is mostly associated with the now popular group Anonymous, its roots go back to the early 1970s with the emergence of social movements (Touraine 1998). The failure of radical Marxism and the exhaustion of the class struggle rhetoric were both partly caused by the beginning of oppositional struggles that were no longer solely addressed at whoever controlled the production of wealth. At this juncture, it is important to highlight the deep controversies of the 1970s, when at the height of the Cold War, Marxist dreams of the USSR were shattered by Aleksandr Solzhenitsyn's book *The Gulag Archipelago* and when the American dream was tarnished by the Vietnam War and the Watergate scandal (Solzhenitsyn 2007). The 1970s could therefore represent the grey area where well-defined conceptualizations of reality came to be reshuffled and re-discussed, and when "the previously assumed centres of politics, the previously dominant forms of struggle and ways of defining struggle, began to fall away to be replaced" (Jordan and Taylor 1994).

However, in these times of change, there is also the technological turn, or aspect, that has to be taken into consideration. The spread of information technologies at the university and military level gave origin to project ARAPANET and to academic networks; at the same time, it gave rise to experimental operations such as the Massachusetts Institute of Technology (MIT) *hacktivities* on mainframes. Finally, with the introduction of personal computers in the 1970s, questions began to arise concerning hardware and software ownership, then still in its infant stage (Jordan and Taylor 2004; Zittrain 2009). Although hacking developed and reached its peak of notoriety in the 1980s and 1990s, the 1970s are nevertheless fundamental for the purposes of this chapter. It is during this time, that the concept of an active public or community which is based on specific issues and that relied on the architecture of hardware of the time (i.e., the mainframe) was formed. With the passing of time, the history of hacking shifted from hardware to software, following some well-known examples, including the electronic hippies, Network Liberty Alliance (NLA) and the Hackbloc Collective with their tagline "hack code not people." From the 1980s onwards, hardware hacking gave way to the software hacking, which focused especially on the code behind the then nascent world of software.

On many occasions anarchist ideas and theories have been applied to hacktivism attempting to define it as a form of civil disobedience. In 1998, Timothy May explained the proximity of hacktivism to anarchism in his "Crypto Anarchist Manifesto." In his manifesto, he used the term "crypto anarchy" as the means that would "allow national secrets to be traded freely and [would allow] illicit and stolen materials to be traded. By consequence, an anonymous computerized market would even make it possible for abhorrent markets for assassinations and extortions to emerge. This would result in various criminal and foreign elements being active users of the CryptoNet." In May's idea crypto anarchy would have created a "liquid market for any and all material which can be put into words and pictures." Therefore, it could be argued that in its many formulations, the association between anarchism and technology has mainly been intended in terms of opposition to the current world order, or what Jordan and Taylor have defined the "neoliberal globalization," put in place since the 1970s and exaggerated in relevance and action after the fall of the USSR in the late 1980s (Jordan and Taylor 2004).

Although hacking as a political force of disobedience should be intended as an antineoliberal force that puts in place tactics of disobedience and criticizes a definite status quo. It is also important to notice that such association is not unique and represents everything but a constant of hacktivist groups. At the core of hacktivism are freedom of expression, a headless organization made of loosely affiliated (if not completely unknown and unrelated) members whose only reason to come together is to address an issue deemed relevant and worth addressing and solving (sometimes skipping the discussion phase). When seen through a Bakunian lens, hacktivism rejects categorizations and fixed theoretical frameworks. Hacktivists as Anonymous or Telecomix (before its blackout) adopt a purely digital methodology of disobedience, reshuffling the cards of control and freedom, both of which are no longer sitting within traditional institutions but are now based on the code. As previously highlighted, the code becomes the new weapon of action, reaction, and repression. Hackers are digital entities that cannot be encapsulated in a specific religious or political credo, rather they become political even when they react against governmental decisions such as that of the "shut down" of the Internet in Egypt in 2011.

HACKTIVISM IS ANONYMOUS

As previously stated, Anonymous is perhaps the best-known modern hacktivist group, along with the now defunct LulzSec. Characteristics of Anonymous are its public "face," the Guy Fawkes' mask, and its very public statements usually entrusted to YouTube and uploaded using untraceable IP addresses or locations. In the past years, Anonymous was involved in some high-profile campaigns, among which were Project Chanology in 2008 and Operation Payback in 2010. In the former case, Anonymous's target was the Church of Scientology, which was seen by the group as attempting to promote Internet censorship. In the second case, Anonymous initiated retaliation against companies and organizations that supported copyright enforcement on the Internet as well as anti-piracy practices. Anonymous, however, has not solely targeted corporations or large organizations. For example, in Canada the group initiated a campaign against cyber-bullying following the highly publicised suicides of Amanda Todd and Rehtaeh Parsons.

The purpose of this chapter is not to delve deeper into Anonymous or its specific activities. This would be beyond the scope of this research as it would lead to an in-depth analysis of this specific hacktivist group, as opposed to the real purpose of this paper which is to analyze the activities and the statements offered by Anonymous and Telecomix during the Tunisian and Egyptian uprisings in 2011. In order to accomplish this, the focus of this chapter will be on the evolution of Anonymous throughout its statements, videos and actions with regards to Project Chanology, Operation Payback, and the events in Tunisia and Egypt.

Chief among Anonymous's characteristics are its boldly phrased statements like: "We are Anonymous, we are legion, we don't forgive, we don't forget. Expect us." Its logo is a simple design of a businessman, whose head has been replaced with a question mark, indicating a headless organization, formed by a loose group of computer savvy people (although this is changing greatly). Although untraceable, the many videos available online are narrated by an atone and robotic voice, and their activities are always aimed at guaranteeing "freedom." However, the concept of freedom as conceived by hacktivists deserves attention and will be discussed later on in this chapter in relation to Bakunin.

Olga Goriunova argues that Anonymous didn't start as a political group but as a headless and quite loose movement with no real purpose though capable of being "nasty." As Gabriella Coleman highlights in fact, the real political turn from hackers to hacktivists happened in 2008 with Project Chanology, the campaign against the Church of Scientology. However, it would be wrong to depict Anonymous as just yet another group of hackers because, as the group itself has stated, "Anonymous is a collective of individuals united by an awareness that someone must do the right thing, that someone must bring light to the darkness, that someone must open the eyes of a public that has slumbered for far too long." In other words this highlights their intent to defend something deemed sacrosanct and worth defending, such as rights and freedom of information. The group also tries to dismantle the myth of the nerd, as someone hidden in the basement with more than one computer screen. As Anonymous itself highlights, among our numbers you will find individuals from all walks of life—lawyers, parents, IT professionals, members of law enforcement, college students, veterinary technicians and more. Anonymous is everyone and everywhere."

What recalls Bakunin's idea of the loosely affiliated voluntary militia, willing to take to war in case of threats to freedom and liberty, is the clear definition of anonymous as a loosely and horizontally affiliated group, one with "no leaders, no single entity" but inspired only by the "collective outrage of individuals." These words first circulated online January 21, 2008 at the beginning of Project Chanology, which was not a random attack but a very focused and precise attempt to dismantle an organization deemed devious and dishonest. Anonymous released videos, comments, and statements supporting freedom of religion but accusing Scientology of having a malign influence on "those who trust" it. Such malignity, Anonymous continued, would only mean the organisation "should be destroyed." The project had an enormous response with online and offline protests organized all over the world and distributed denial of service (DDoS) attacks to the center of Scientology in the United States, as well as bombarding the centre with prank calls and faxes.

In very similar fashion, Anonymous launched Operation Payback aimed at pro-copyright and anti-piracy groups and organizations. The campaign was initiated on 4chan.org, an anonymous image board, as a response to DDoS attacks carried out by Aiplex Software against torrent sharing sites on behalf of major media industries. Most notable among these was the DDoS attack on The Pirate Bay which shut down the website, and other noncompliant torrents were denounced and reported shortly after. As a response 4chan organized a counter DDoS at the expense of randomly chosen corporations such as the Motion Picture Association of America (MPAA) and the Recording Industry Association of America (RIAA). The targeted organizations' websites were down for only a couple of hours, but in that short amount of time approximately one-third of the encrypted and sealed information of the RIAA went public, through magnets and hence shared online.

Although these two examples highlight the particular antineoliberal character of Anonymous, the group's anticyber-bullying activities by contrast have little or nothing to do with a political order or status quo but rather with ethics and the freedom of online activities. More specifically, Anonymous advocates freedom of activity and movement online, and the ability to do so without being harassed and blackmailed. In taking this position, the group unwittingly policed the Internet, giving more space and more credit to what Telecomix hacktivists defined a "do-ocracy."

Somewhere in the grey area between antineoliberal and nonpolitical activities lies Operation Assange. Though it has been primarily categorized as antineoliberal, a deeper analysis reveals that there is more to this campaign than this. Though political in nature, the campaign stemmed from the indignation of a group that saw the hypocrisy of certain governments, corporations, and organizations when it came to the case of Julian Assange, the founder and main editor of WikiLeaks. In December 2010, when the operation was launched, the initial goal was to retaliate against those trying to stop WikiLeaks from publishing classified US diplomatic cables. However, as controversies grew around WikiLeaks and Assange, the campaign began to target both actors that have been in favour of more restrictive actions controls of the Internet as well as those that have promised changes and more progressive actions.

Reflecting on the previous example leads us to question whether these groups are purely political, or if they fall into some other categorization. On many occasions Anonymous has stated that "knowledge is free" without adding any additional explanation or statement, or alluding to any particular actor. As a result, when groups like Anonymous and Telecomix talk about freedom, it is no longer possible to categorize these statements as political in an anti neoliberal sense. Instead, their political nature becomes political because of a purely anarchical view focused on freedom, including that of obtaining, accessing, and sharing information online. For Bakunin, the concept of freedom is intended as liberation from domination, form rules, and norms imposed from above. The concept of freedom as promoted by Anonymous, and shared by many other hacktivist groups, is concerned with freedom of expression and freedom to "know," effectively resting on the idea of liberation from norms and rules that aim at modelling and filtering the access and consumption of information.

FROM THE ARAB SPRING TO CODE: ANONYMOUS, TELECOMIX, AND THE DEFENCE OF FREEDOM

In an interview to *OpenSource*, an online publication and community, Peter Fein, a spokesperson of the Telecomix group, highlighted the loose and anonymous relationship between the group's members, whose main ideals are freedom of information and "do-ocracy." This latter term was used by Fein to describe the spirit of hacktivists during the Egyptian Revolution in 2011, and whose efforts were seen as effective because of their ability to "just get things done."

On January 27, 2011 news circulated online with regards to a "Facebook shut down" at the hands of the Egyptian government (Whittacker 2011). The decision seems to have been prompted by the mass street demonstrations of January 25 and the incessant calls for an end to President Muhammmad Hosni El Sayed Mubarak's regime, as well as a tactical use of social media tools such as Twitter and Facebook.[1] What happened touches the legal and technical aspect of the Internet infrastructure. On the legal aspect, in Egypt, like in many other restrictive countries, telecommunication companies negotiate licensing agreements with the governments in order to get into the market. In this case, one of such restrictive licensing agreements was to comply with any requests coming from the government to limit the Internet traffic in the country (Freedom House 2011). In this case, it could be said that the Internet service providers (ISPs) in Egypt acted upon the functioning of the previously introduced BGP tables, fundamental for any routing decision of the Internet, thus restricting what could be called the "communication funnel" (Cisco Management Systems 2012).[2] In Egypt, the Mubarak regime had the possibility to "shut down" the Internet overriding the BGPs through the ISPs, which in this case acted as network administrators. As a result, Noor Data Networks was the sole provider from a list of ten different ISPs to maintain its access to the Internet infrastructure during this time.[3]

By consequence of these actions, Internet traffic originating from and leaving Egypt dropped by approximately 88%, meaning that government's "rearguard battle" (Shirky 2009; 2011) was not as desperate as one would have thought (Williams 2011). This shutdown had served to further motivate the tech-savvy population with the support of hacktivist group Telecomix who in some instances used "ham" radios (that are usually created by amateurs for entertainment, or experimental purposes), to help keep the movements and street protests alive.

Egypt owns a relatively small number of autonomous systems, which are the networks under a single administrative control that share the same information, and two Internet exchange points (IXPs) based in Cairo, CR-IX and CA-IR. The two latter are open only to the licensed ISPs with international connectivity. However, because Egypt's IXPs are fundamental for keeping an Internet presence in Africa, the government could not physically intervene on them for political and tactical reasons. Instead, this interference took place on the line of code, or more specifically, on the BGP tables that restricted traffic coming from external and foreign sources. This way, when information from a "foreign" source like Google attempted to reach its destination in Egypt, the BGP tables failed to recognize it and did not pass the information along as would normally happen. Therefore, Egypt was cut off from the greater Internet, and foreign services such as Google, Facebook, and Twitter were not accessible.

As a result, algorithms, code, and routing protocols became the main actors around which the whole series of tactical actions deployed by street demonstrators and hacktivists to overcome the Internet disruption centered upon. The issues of the political, social, and economic discontent became enriched by limitations to access information, communication and freedom of speech through social networking sites, Facebook, and Twitter playing the leading roles.

In response, Telecomix immediately undertook tactics to restore the flow of communication, responding to a very Bakunian idea of revolution and freedom. Once again, these tactics took place in the line of code and put in place in the name of "do-ocracy." This intervention by hacktivists in fact served the purposes of guaranteeing communication during the revolution, even before the actual Internet blackout. While the Internet was still "up," the group helped building mirrors and proxies, respectively duplicating entire datasets of websites in order to provide the same information to multiple points at a faster speed (Cisco Management Systems 2012), with servers acting as an intermediary between an Internet user and his destination website. Telecomix also used Internet relay chats (IRCs), as well as a manual relay to Twitter, to allow communication access to millions of Egyptians who could not access the social networking site. After the shut down, Telecomix supported the set-up of dial-up modem lines, whose number was circulated on Twitter as +46 850 00 99 90. This information could be found on Telecomix's Twitter page and under the #Egypt hashtag. The message carrying the telephone number and the password to access the dial-up modem was followed by the message "Only for Egypt, Respect That Please!."[4]

The tactics put in place to avert the "shutdown" also included the establishment of the so-called hams, or amateur radios, which in this case were used for emergency purposes. A video posted on Vimeo by Telecomix, titled "Egypt Hamradio Op," shows the code used to install and establish radio communication inside and outside the country.

Therefore, with a very effective and guerrilla style message, Telecomix claims that "When countries block, we (d)evolve. We are Telecomix and we build" (Fein 2011; Noirfalisse 2011). This last sentence can be linked to Fein's assertion that whereas Anonymous destroys, for example by using DDoS attacks, Telecomix builds or rebuilds.[5]

Telecomix's hacktivities in Egypt followed Anonymous's involvement in the Tunisian Revolution, also known as "Operation Tunisia." In a press release posted on YouTube, Anonymous claimed that since "the Tunisian government has decided [...] to control this present with falsehoods and misinformation and restrict the freedoms of their own people" the government "has made itself an enemy of Anonymous." In this case, though Telecomix's tactics had been seen as less malign that those of Anonymous, both groups' intentions are aligned in that they seek freedom from the control of information and act in the name of free speech. Therefore, next to the traditional DDoS attacks to governmental sites, Anonymous circulated information on how to install Greasemonkey (an application extension for Mozilla Firefox that enables scripts[6] to be added to the bowser thus allowing it to change the content of a page or a particular online document) and TOR, an anonymizer free software to download along with Vidalia in order to access safely the

Web and hide the IP address (unlike other HMA for example, TOR doesn't allow the user to have control over the IP location).

CONCLUSION

The introduction of hacktivism as a political force, and the subsequent analysis of Anonymous's and Telecomix's activities, shows how Bakunin's thought of destruction of the state can be applied to the modern world order. By reviewing Bakunin's idea of fight and violence, and adapting his concept of the voluntary militia and defence of freedom to the current episodes of hacktivism, it is possible to highlight the inner anarchical nature of hacktivism. However, this does not only regard their destructive aspects or their antineoliberal ideals, but also their defence of freedom and their effort to understand and embrace local issues. What this chapter has addressed as the *hacktivities* performed by Anonymous in Tunisia and Telecomix in Egypt are not to be read solely in relation to their detrimental effects on the reputation of governments. Such *hacktivities* have an anarchical value because they have been applied recognizing the universal principles of freedom and the very local nature of the issues reclaimed and vindicated by the masses. Through the use of code, identified as the weapon of choice in these times of digital transformations, the very local issue of people marching on the street has been addressed with the help of digital media, and a digital commune has been created around the events of January 27. Therefore, despite the attempts of governments and regimes to silence the demonstrations and interrupt communication between the demonstrators, alternative guerrilla style tactics came to the fore as a weapon of resistance. It is here that low-tech radios made by amateurs, in addition to the use of Greasemonkey and TOR, have given way to the anarchical social revolution as envisioned by Bakunin in the wake of the Paris Commune in 1871. As it has been previously discussed, anarchism is what Bakunin referred to as a headless assemblage of people, masses from very different social classes, whose main intention is to change a status quo in the name of freedom.

The anarchy Bakunin envisioned was the destruction of the state as an apparatus of norms, directions, and regulations applied over the masses and not recognized (nor accepted) by them. His thought of a social revolution brought to life by the spontaneous association of the masses, recognizing the local nature of their issues, serves the purposes of contemporary history. Weapons of defense and attack are now concentrated in lines of code, scripts, and APIs that can merge topics, aggregate publics, and stir debates. The social revolution as envisioned by Bakunin is effectively put in place though lines of code with a voluntary militia willing to arm itself with digital weapons in case of threat of the commune. In line with Bakunin's thought, it could be additionally argued that this voluntary militia can embarrass governments with the publication of inconvenient information, ridicule them and join the masses in their quest for change (here not intended in recent electoral tag lines). Readapting Bakunin's thought of freedom and association, his definition of voluntary militia deserves attention and reflection. In *Revolutionary Cathechism,* Bakunin claimed that "all able-bodied citizens should, if necessary, take up arms to defend their homes and their freedom. Each country's military defence and equipment should be organized locally by the commune, or provincially," offering Switzerland as an example. What this chapter has shown is how Anonymous and Telecomix, have acted as the voluntary militia and how they have "taken up arms to defend" freedom. As previously highlighted, the concept of violence is here redefined and finds new life. Additionally, earlier in the chapter Bakunin's concept of destruction had been revisited highlighting its non-nihilist nature. In consequence, it has been the intention of the chapter to re-examine Bakunin's ideas in a more modern context potentially opening the door to new theories or academic discussions. Therefore, what Anonymous, Telecomix and Lulzsec do, or have done, falls outside the classical Marxist and post-Marxist categories of class struggle—a seemingly fair, although highly organized and rigid, revolutionary socialism developed through the "non-political or anti-political social power of the working classes in city and country, including all men of goodwill from the upper classes who break with their past and wish openly to join, them and accept their revolutionary program in full" (Dolgoff 1972).

Finally, as it has been argued, hacktivists and their hacktivities almost always argue for the defense of some form of freedom. More specifically, freedom of speech could be recognized as the universal ideal that drives hakctivist, which is a wider belief supported by local issues. More than ever the centrality of the local issue is at the core of these hacktivist groups who decide to intervene looking at the specific case, as well as at the local nature of the issue, which is not limited to regime change or revolutions. Therefore, while Telecomix reflected on the economic, political, and social nature of the Egyptian Revolution, Anonymous learned from the Tunisian people's resistance against their government's attempts to control and spy on them. In both cases, the universal ideal was freedom, which depending on the local context of each country, became fused with local causes that had materialized in response to the restriction of these freedoms. As a result, this lacking freedom found its expression through, at times destructive, outlets, as was the case with code and various Internet applications and social networks. By consequence, the state that Bakunin had envisioned is also changed in nature in that it now attempts to control and exert power over the code in an attempt to establish a new weapon of control, power, and repression. In the hands of hacktivists, the code became a new source of counterattack for the commune (communities as envisioned by Bakunin). The anarchist thought as seen by Bakunin serves the purposes of better studying the world of social networking sites and online networks, especially when employed to follow or promote a particular cause. Finally, as a provocative conclusion to this chapter, it could be hinted that Bakunin's ideas, especially when applied to the analysis of digital media tools and the recent wave of Revolutions 2.0, may help establish a model for interpreting bottom-up changes, grassroots movements, and democracy-from-below in these changing digital times where socialist utopias that have been so "easily appropriated by capitalism."

NOTES

1. Available at https://twitter.com/telecomix/status/31065485730652160. Last accessed June 21, 2013.
2. An Internet exchange point (IXP) is the physical infrastructure that allows the Internet service providers to access the Internetwork and exchange traffic between their autonomous systems.
3. Noor Data Networks is fundamental for the functioning of the Egyptian Stock Exchange.
4. Available at https://twitter.com/telecomix/status/31065485730652160. Last accessed June 21, 2013.
5. Video available at http://opensource.com/life/11/6/telecomix-anonymous-anarchy-and-getting-things-done-through-do-ocracy. Last accessed June 21, 2013.
6. The IP is responsible for the routing and address of datagrams (packets of information) from a host to a destination.

REFERENCES

Brener, A. and B. Schurz. 2009. "Anti Technology of Resistance." *Subsol Anarchist Online Journal*. http://subsol.c3.hu/subsol_2/contributors/brenertext.html

Chun, W. 2006. *Control and Freedom. Paranoia in the Age of Fiber Optics*. Cambridge: MIT University Press.

Cisco Management Systems. 2012. *Cisco Handbook of Technology*. http://www.cisco.com/en/US/tech/tk365/technologies_tech_note09186a00800c95bb.shtml#howbgpwork

Collins, P.H. 2004. *Dictionary of Politics and Government*. London: Bloomsbury Publishing.

Critical Art Ensemble. 1994. *The Electronic Disturbance*. Brooklyn: Autonomedia.

Deleuze, G. 1992. "Postscripts on Societies of Control." *L'Autre Journal* 1.

Dolgoff, S. 1972. *Bakunin on Anarchy: Selected Works by the Activist-Founder of World Anarchism*. New York: Knopf.

Fein, P. 2011. "Interview by R. Shuherle, *Telecomix, Anonymous, anarchy, and getting things done through the do-ocracy*." OpenSource. http://opensource.com/life/11/6/telecomix-anonymous-anarchy-and-getting-things-done-through-do-ocracy

Foucault, M. 1995. *Discipline and Punish. The Birth of the Prison*. New York: Vintage Books.
Freedom House. 2012. *Egypt*. http://www.freedomhouse.org/report/freedom-net/2012/egypt
Jordan, T. and P. A. Taylor. 2004. *Hacktivism and Cyberwars. Rebels with a Cause?* London: Routledge.
Latour, B. 1999. *Pandora's Hope. Essays on the Reality of Science Studies*. Cambridge: Harvard University Press.
Law, J. 1986. *Power, Action and Belief: A New Sociology of Knowledge?* London: Routledge & Kegan Paul.
Lehning, A. 1973. *The Selected Writings of Mikhail Bakunin*. London: Cape.
Maximoff, G.P. 1953. *The Political Philosophy of Mikhail Bakunin*. New York: Free Press of Glencoe.
Morris, B. 1993. *Bakunin: A Philosphy of Freedom*. London: Black Rose Book.
Noirfalisse, Q. 2011. *Telecomix: Hacking for Freedom*. OWNI.eu. http://owni.eu/2011/08/01/telecomix-hacking-for-freedom/
Shirky, C. 2009. "How Media Can Change History." *TED Talks*. http://www.ted.com/talks/clay_shirky_how_cellphones_twitter_facebook_can_make_history.html
Shirky, C. 2011. "The Political Power of Social Media. Technology, the Public Sphere and Political Change." *Foreign Affairs* 1. http://www.foreignaffairs.com/articles/67038/clay-shirky/the-political-power-of-social-media
Solzhenitsyn, A. 2007. *The Gulag Archipelago 1918–1956*. New York: Harper Perennial Modern Classics.
Touraine, L. 1998. "Social Transformations of the Twentieth Century." *International Social Science Journal* 50, 165–171.
Williams, C. 2011. "How Egypt Shut Down the Internet." *Think Insights*. http://www.thinkwithgoogle.com/insights/library/studies/our-mobile-planet-Eygpt/
Witthaker, Z. 2011. "Egypt 'Shuts Down Internet' Amid Further Protests: Facebook Web Traffic Drops." *ZDNet*. http://www.zdnet.com/blog/igeneration/egypt-shuts-down-internet-amid-further-protests-facebook-web-traffic-drops/7915
Zittrain, J. 2009. *The Future of the Internet: And How to Stop It*. London: Penguin Books.

ONLINE SOURCES

Anonymous. (Producer). (2008). *Project Chanology*. http://www.youtube.com/watch?v=JCbKv9yiLiQ
Anonymous. (Producer). (2010). *Operation Payback*. https://www.youtube.com/watch?v=2Tm7UKo4IBc
Anonymous. (Producer). (2010). *Operation Tunisia*. http://www.youtube.com/watch?v=BFLaBRk9wY0
Telecomix. (Producer). (2011). *HamRadio Operation*. http://vimeo.com/19688181

Chapter 2

New Media and Empowerment in the *Indignados'* Movement

"If you want that no one else decides for you, so that no one else speaks for you"

Tommaso Gravante and Alice Poma

The Spanish *indignados*' movement that came to public light on May 15, 2011 (shortened to 15M) with protests in many cities throughout the peninsula with the participation of citizens who felt disenchanted by the Spanish democratic model and indignant due to the privileges of the political class, not only represents a novelty in the Iberian political arena, but also it has contributed to reopen the debate regarding the relation between social movements and new technologies. As we will see in this chapter, among the results of this movement it is worth mentioning the opened discussion regarding the legitimacy of the mainstream media. Hence, the protest becomes an experience of learning and empowerment that in the communicational field is manifested by the development of new independent self-managed projects that respond to the need of creating communication from below, as we can notice in the words of a member of Seville's 15M: "Another type of communication is possible by breaking what we have known, that is, breaking this pattern and creating your own, and that it shouldn't look like anything: it should look like you, like your beliefs, whether individual or collective."

In the *indignados*' movement the practice of net activism, which comprises those political experiences nourished by the use of new technologies, acquires a great deal of relevance, although it is important to remember that the movement was not born with said activism. Since the 1990s in Spain, experiences such as *Nodo50*, which was created in 1994, *Sindominio.net* (1999), or *La Haine* (2002), just to name a few, have been pioneers in the process of transforming the Internet and new digital media, turning them from simple communication tools into communicative spaces where Spanish experiences of protest have laid down in the subsequent years (Martin and Domínguez 2006).

In addition, in Spain as in other parts of Europe,[1] net activism characterizes by questioning the dominant relations and criticizing the dominant discourse with different views (Pasquinelli 2002). With the years, these alternative experiences of communication have become political struggles where images of the world of the participants, as well as their needs and desires, are claimed, expressed, and broadcast, as the *indignados*' movement claims; while the access and broadcasting of the information have become core topics in protest movements (Marí Sáez 2004; ROE 2006).

From these reflections, instead of focusing on the movement and the communication strategies used, in this chapter we will present an analysis that shows how the appropriation and use of digital media becomes an antagonistic event, devoting special attention to the experience of the protagonists. We will also demonstrate how the process of appropriation of the social media is linked to the empowerment process that people go through, transforming them into political subjects.

Appropriation and empowerment are two processes developing contemporarily, without being consequential. Therefore, we have chosen to organize the analysis of these processes in three sections that are intertwined. In the first of them we will look at why and how people appropriate the social media; then we will see how this process of appropriation transforms into a process of identification and reflection over communicational media and processes; lastly, we will analyze how this practice of action-reflection is linked to a transformation of awareness and behavior of the people interviewed. Finally, in the conclusions we will jointly reflect on the results of these analyses. In order to attain the objectives of our analysis, however, we will present the following sections, the research methodology and techniques employed, the Spanish context, and some characteristics of the *indignados*' movement, aiming to provide the reader with the necessary tools to comprehend the analysis.

OBJECTIVES AND RESEARCH METHODOLOGY

For the development of this research, in addition to the general knowledge of the movement on a national scale, we have chosen a case study, Seville's camp. This research technique allows the comprehension of the conflict through subjective categories and contextualized knowledge and, helps to explain "the reasons that lead the person to act in a determined way giving attention to the context in which the action develops" (Coller 2005, 20). As well as being supported by ethnographic observation (May 2011 to November 2012) and documentation collection, the analysis is based on in-depth interviews with people who participated in the communication committee during Seville's camp (carried out in October and November 2012), giving privilege to an idiographic approach that considers subjects not as variables, but as a whole within their ecological, social, and historical context (Hernández 2000, 53).

The semistructured interviews have included own strategies of biographical, individual, nondirective, nonstandard, open, guided, and deep interviews (Bogdan and Taylor 1987), and also the episodic interview (Flick 2000 and 2004) that "facilitates the presentation of experiences in a general, comparative way, and at the same time it makes sure that those situations and episodes count in their specificity" (Flick 2004, 119). Centering our analysis on the biographical experience and the emotional dimension of the participants, we could not but choose qualitative research techniques, based on the idea that "qualitative research has a more holistic vision and provides higher importance to social interaction processes" (Della Porta 2010, 13).

Now we will briefly describe the Spanish social context that is inserted in the *indignados*' movement as well as some elements that characterize this social mobilization, to then show how the concept of the social media takes a new meaning based on the subjective experience of people, and adding the determinant elements in the practice of appropriation of these media. We have also decided to enrich our analysis with testimonies extracted from the interviews, since, as Romero and Dalton have written, "the human accounts are more eloquent" (2012, 12).

SPAIN: "ERROR 404: DEMOCRACY NOT FOUND"

The problems and demands that the *indignados* claim, did not arise along with the movement; in fact, as James Scott writes, in each society there is a "hidden transcript" of subordinate people which in special events[2] emerges and makes itself public. This hidden transcript "represents a critique of power spoken behind the back of the dominant" (Scott 1990, xii), it emerges in the social and marginal spaces when there are more people who share it, and "it exists only to the extent it is practiced, articulated, enacted, and disseminated within these offstage social sites" (Scott 1990, 119). Therefore, before introducing our analysis, it is necessary to frame the protest within the country's political and cultural context to be able to understand the importance these experiences can have for the society.

A scarce dialogue between the political class and the citizenship characterizes the Spanish political culture, since in the country's history the absolutism with episodic democratic moments has prevailed (Sierra 2012). In the three decades of the current democratic system, it has been impossible to implement an effective dialogue between citizenship and institutions, and, even worse, some heritage of Franco's military dictatorship such as the monarchy is still preserved, which was never subjected to a popular referendum in contrast to Italy, for example.

Neither the leading political forces of the "democratic transition" in the 1980s nor the successive governments of democracy were interested in developing a public arena whose foundations were plurality and diversity of ideological schools. Rather, they nurtured a restrictive party system, characterized by a concentration of economic, judicial, and political powers, including the media. Needless to say, Spain is characterized by a political system that is essentially founded on a political machine network where dictatorship-related elements have not even been eradicated (Sierra 2012).

The *indignados*' movement is framed in a cycle of mobilizations that emerged since 2008 in Europe, in response to the consequences of the economic and social crisis that has contributed to nurturing, on the one hand, the social discomfort due to the worsening of the quality of life of most citizens; and on the other hand, it has increased the disaffection towards the political class and towards the large lobbies and economic groups in the country, helping to weaken the progressive constitutional state model.

In this context, the administrative elections of May 22, 2011, constituted the perfect stage for the protest of those feeling disenchanted by the Spanish democratic model and indignant at the privileges of the political class. One Sunday before the elections, through a call that travelled mainly online, successful demonstrations were held in more than sixty cities throughout the country. The euphoria from the high participation in the call made the group 15M in Madrid decide, within an assembly, to carry out a camp in Plaza del Sol, inspired by the images and events happening in Plaza Tahrir in Egypt as well. The violent evacuation of the camp in Plaza del Sol in Madrid was the triggering element for the movement to spread and the practice of the camp in many other cities. Next day, a crowd filled Plaza del Sol in Madrid, and other camps were organized in the main cities around the country. With the resulting violent evacuations of the camp in Barcelona and other cities, the movement became stronger, finding wide support in citizens who were already beaten by the economic crisis and generally tired due to the abuses of the political class.

From these premises, in the next section we will describe some characteristics of the movement that can help us comprehend how their values influenced the appropriation and use of the social media.

THE *INDIGNADOS'* MOVEMENT: "DEMOCRACY UPDATED 2.0"

Despite some slogans and the common sentiment of indignation, the demonstrations and camps of 15M throughout the country were different in terms of their practices and strategies. The assembly experiences were in some cases strengthened in the neighborhoods of urban centers while in others the movement spread towards peripheral towns; in other cases the assemblies were oriented to concrete struggles such as the struggle against evictions.[3] Undoubtedly, all these experiences were characterized by two elements that belong to the movement: the participation of ordinary people and the development of an own informative system linked to the use of the social media above all.

The participation of the ordinary people has been possible because in the Spanish society the institutional left wing does not have the power of mobilization that other European countries do; in contrast, the relations that are built next to each other are determinant, where solidarity and social and affective relations have a key role among people, and have strengthened thanks to the economic crises. In this context, the traditional channels of communication like the word of mouth were then accompanied by the appropriation by citizens of the new communication technologies that allowed higher interaction, the horizontal discussion and free exchange of information, favoring the self-organization and coordination of the movement.

The generalized mistrust in the political class has also caused the movement to rejects the participation of any political organization from the very beginning, banning flags in their protests and accepting the individual collaboration rather than corporate. The participants, instead of activists, are an essential component of the movement, providing their voluntary collaboration driven by ethical and emotional motivations instead of material incentives that respond to a cost-benefit or political machine logic.

Among the repertoires of the protest, in addition to the camps, we can find marches, street or virtual sit-ins, flash mobs, artistic improvisations, and currently, the *escraches*.[4] The spontaneous creation of camps can be considered as a consequence of the will of citizens to recover spaces and decision capacity, though the process has not been easy since the population is generally accustomed to delegating important elements of their lives, such as their children's instruction, their health, and so on, therefore self-organization has a necessary learning process, as this interviewee states: "It is a whole learning process. Now everything works, but it's like an organized disorder."

Some other characteristics of this movement are heterogeneity, the diffuse participation and local character of camps that have given the movement creativity, irony and resilience, since the strategies adapt to the movement and territory, or as said by an interviewee: "15M is a whole of diversity that cannot be generalized . . . There are many diversities and sensitivities." The *indignados*' movement is also characterized by its horizontality and decentralization, which is the reason why the general, neighborhood, commission, working group assemblies, and so on represent the core of the movement itself. The emphasis given to the participation of individuals in the activities of the movement and in the autonomy of camps arises from the necessity of real, democratic, and direct participation on the part of the citizen in the political life. This practice of democratic and local democracy not only characterizes the movement, but also strengthens and legitimizes its discourse, evidencing the failure of the representative democratic system that excludes citizens from the decision-making processes even more.

If we consider identity as a process (Della Porta and Diani 1997, 109), for the past years the 15M movement has been constructing its antagonistic identity between "us" (the people, as a people, the movement, etc.) and "them" (politicians, bankers, business people, etc.) through ritual forms, narratives, performances, and so on. Identity is strengthened by sharing personal experiences, learning about foreign people's stories, with different struggles that could be transmitted thanks to new communication technologies that have been determinant. As well, identity has its roots in the local dimension, that is to say, in the squares of every place, city, or village, which become the point of reference for the citizens who want to participate and get involved but who do not know anybody who is engaged in the movement. In conclusion, in relation to the motivation for action and participation, we can assert that in the case of the *indignados*, said motivations are individual, there is not a corporate call to participate; it is the citizens who feel they want to do something, as this woman has expressed: "I didn't want to be quiet because to me this was important. To me everything was so big-hearted that I wanted to take part in it." This was a brief description of the movement. We believe it is enough to understand the analysis proposed. In the next paragraphs we will now proceed to show how an alternative medium of communication is redefined from the perspective of the subject and how the process of appropriation and use of the social media is re-signified from the experience of people.

ALTERNATIVE MEDIA: A SPACE FOR INDIVIDUAL AND COLLECTIVE TRANSFORMATION

Our analysis of appropriation and alternative use of the social media in terms of the *indignados*' movement assumes that these experiences of communication cannot be considered as mere technological broadcasting channels of alternative information in contrast to mainstream media production. From our point of view, these media experiences are spaces where, irrespective of the capability to broadcast and the most evident results that the produced information may have, they influence change in people and their transformation into political subjects, and consequently, a social change at a wider level.

One of the first characteristics that emerges from our analysis is that these experiences reflect and express the everyday nature of people since they are directly related to people's needs whether ordinary (work, health, education, social rights, or national identity claims, etc.) or extraordinary (the outbreak of a conflict, the defense against a threat or repression, etc.). In this process of appropriation, people give a sense of belonging to media tools; they value them and learn how to use them to meet their own needs and interests or those of their social group. Therefore, the use of the social media is essentially linked to the social and cultural body, and it is from this perspective that we propose our analysis.

Looking at the social media from the subjects' point of view, it is observed that these experiences are influenced by the *Do It Yourself* (DIY) culture (McKay 1998; Spencer 2008), which erodes the dependency on professionals and intellectual elites, who are substituted by autodidact, informal practices, and collective abilities. The DIY practice transforms into an educational process experienced by both the protagonists and the community to which the medium is directed, since the audience is always about to participate in the media experience. This relation of horizontality between the sender and the receiver allows the alternative medium to directly involve the audience in the production and distribution of its own messages, thus avoiding the commercialization of the characteristic contents of the mass media.

In our case of study, the practice of self-publication made the narratives, desires, and claims of the movement be constructed by the very same participants in the social media used by the *indignados.* The participants constructed a collective identity they could feel identified with using transgressive practices, such as irony, humor, culture jamming, provocation, and fake. The participants of the movement, through the social media, constructed their own interpretative frame of power and politics, challenging the country's dominant political structures and putting the vertical forms of political, union, and social organization in check.

To sum up, from these elements it is evident that for a better comprehension of the process of appropriation of the social media in a social protest, it is necessary to carry out an analysis of the cultural and biographical dimension of the protest, rather than just exclusively rely on the knowledge of the level of people's technical culture (Proulx 2004) or the distribution of the information capital (Hamelink 1999). Inverting the look towards alternative media from the subject's point of view involves reinserting the process of appropriation of the social media in the social construction of meanings given by the protagonists of the media practice with their motivations and imageries, or in Michel De Certeau's words "the patchwork of everyday life" (2000, xlix).

Having these reflections in mind, in the following paragraph we will present the main elements that have constituted this patchwork in the *indignados'* movement regarding the appropriation and use of the social media; we will go on to describe the process of empowerment that people have lived. As stated in the introduction of this paper, in the next section of our analysis we will discuss why and how participants of Seville's 15M have constructed their own channels of alternative communication.

SOCIAL MEDIA: DO IT OURSELVES

In the same way that motivations of participants in social mobilizations are not limited to material or initial objects of the protest, being as well nourished by a hidden transcript which is always present in society (Scott 1990), as was the case, for example, of mobilizations in defense of the Gezi Park in Istanbul in 2013, the appropriation of new media must not be seen only in terms of the availability of resources or as an exclusively technical issue. As we have seen before, in order to deeply understand the process of appropriation of the media we will have to consider the context in which the medium develops, the individual and collective identity that the medium presents for its community of reference. In other words, it is necessary to see the social media not as a technical communication tool, but instead, as a cultural mediation through which people know each other again and make their own social imagery emerge (Martín-Barbero 1987). Due to these reasons, the appropriation of alternative media arises as a response to the demands that come with the movement. The necessity to communicate what was happening pushed people toward these communication tools regardless of their technical knowledge. Appropriating something new was just the first step, which allowed them to put down the fear of the unknown, as this interviewee says: I had no experience in communication but I wanted to give something with my work.

This appropriation then arises from the desire to participate in the movement and does not require any previous knowledge of technologies and communication techniques. Despite this, learning to use the media and sorting out technical, economic, and organizational problems represents the next step in the process of appropriation which, as this person said, is not so immediately and easily solved: "At the beginning it was chaos, there was no light, we didn't have Internet, we used anyone's networks." It is interesting to observe that in the process of appropriation the practice of sharing knowledge and resources is substantial to overcome the digital gaps since it was thanks to many people who offered their abilities that technical problems could be solved and projects could be carried out and knowledge could be passed down, as one interviewee mentions: "The lack of technical knowledge was a minimum problem."

In 15M, sharing resources and knowledge, by both regular citizens who made their WiFi public, or experts and activists who shared their experience, was a common practice in the camps and is one of the strengths of the movement. An outstanding example is the server named tomalaplaza.net, which was available from Madrid to host pages of the various camps. In fact, many people went to the capital city to share, learn, inform, and get informed, raise questions, look for solutions, and so on, and go back to their own camp with all this new knowledge in order to strengthen it. But, in spite of this tie with the camp in Plaza del Sol in Madrid, all the other experiences have been independent from the beginning, and each of them started to work by taking advantage of the human and technological availability in their local community. In Seville, as an interviewee told us, people gathering in the assembly felt from the first moment the necessity to produce their own medium of communication independently of the camp's page in Plaza del Sol: "We created a blog the first night we were camping but within a week and a half with the space in the new server we moved and changed to Word Press."

Regarding the communicative tools used in camps we have been able to confirm there were many: from blogs to social network platforms, from streaming to an analogical radio program. Each technology constituted a different possibility for communication, targeting different audiences, based on the demands of people that were in charge of the medium; in the words of the protagonists: "Each channel of communication has its nuances, so I don't think ones are better than others, I believe all of them are necessary and possible because people are different and each medium has its own register, a different way of communication that makes people feel more comfortable." This communicational experience provided people who contributed to constructing alternative media of the movement with new arguments and new capabilities in such a way that the interviewees stated that each of the tools employed had a different use, as we can read in the following testimony: "It was an experience that made me see how Twitter means much more in terms of teamwork rather than a blog or Facebook. A blog is more like reporting, but if what you want is learning about the breaking news, if you really want to know what is happening, Twitter is more immediate." From what we have described so far, it is observed that the process of appropriation and use of digital technical tools is the result of collective abilities, of making something common, of discussing objectives and sharing knowledge among the community that is created while surrounding the alternative medium. In this collective learning both the means and the ends are discussed and decided collectively. This process, essentially collective and shared, substitutes the positivist individualism of DIY as a result of the collective practice of sharing, that is to say, due to decentralizing knowledge where innovation of practices, meanings, and sometimes tools takes place. In short, we can assert that the practice of DIY was transformed into an educational process experienced and lived by the protagonists and that evolves from Do It Yourself to Do It Ourselves.

As we have described in this section, the practice of appropriation of the social media in a protest involves many more aspects than just the technical one, and in fact, it does not end with the use of the media, but new elements take place, such as the identification of people with the media and the process of action and reflection about their use, as we will demonstrate in the following section.

FROM 'WE DO THE MEDIA' TO 'WE ARE THE MEDIA'

As aforementioned in the previous section, the practice of Do It Ourselves at first takes people to interact with the medium to modify it, adapt it, and provide it with meaning; an example of all this is the name of the domain itself: *Toma la plaza* (Take the square). The name on the Web is the continuation of the actions happening in the streets and is the continuation of the collective imagery that sustains the real camps transformed into public squares for discussion. With the construction of the *Toma la plaza* portal, the camps became hybrid spaces where participants could experience other forms of doing politics in relation to proposals that were developed and discussed in the real camps as well as on the virtual applications.

Subsequently, as a second stage, the participants of the movement got identified with the communication tools, and the social media were absorbed as part of the movement's collective identity, becoming channels to put into practice core values for the movement such as transparency in decision making, horizontality, and solidarity. And in fact, the new technologies did their job as they allowed broadcasting the practices and the imagery of a new political system vindicated by the movement and founded on the idea of a direct democracy.

In parallel with the process of identification with the medium, people start and develop their own interpretative and critical frame of the use that must be given to the media, turning the practice of appropriation into an action linked to reflection (Freire 1970; 1980). In Seville's camp the participants, after providing a medium of their own and sorting out the logistic and technical problems related to infrastructure and media, expressed in the assembly of the communication committee the claim of providing a guide that reflected the decisions agreed in terms of what type of news to broadcast and how it must be published. These needs materialized in the creation of the protocol on communication that, consistently with the values claimed by the movement, was written and approved with the consensus of all, as we can read in the following testimony: "We passed the protocol to the Neighborhoods and Peoples Assembly, and it was agreed by consensus, by the whole assembly, so it became the official protocol. And whoever joined the communication committee already had the basics of how to work and some dynamics that we had already agreed by consensus." The practice of appropriation involves an action-identification-reflection process that includes the analysis the protagonists carry out in terms of technology, which is not associated directly to social change and to which absolute trust is not given.

Instead, they recognize themselves the limits of society digitalization that excludes the majority of the citizens who do not have access to the Net. These are the reasons why the people interviewed, despite recognizing the importance of the social media in their movement, defend the need for not moving away from their own local community of reference, as this interviewee has expressed: "There are many people like us who have bet to digital, and digital technologies have been fundamental for this process [el 15M], but we cannot abandon the streets because there are a lot of people who still don't have access to Internet or don't know how to use these tools." This testimony allows us to understand why the people interviewed who participated in the communication committee defend the traditional methods of transmitting information such as the word of mouth and social relations, despite being the first in transmitting information through the social media. Not surprisingly, as one interviewee stated, in the climax of the digital era, the local camps of 15M chose to combine the social media with the traditional mechanisms of communication, demonstrating, once again, consistence in their decentralized and horizontal organization: "The medium that gets faster to people on foot is the word of mouth, it's the streets. I still believe that the streets are the greatest medium to transmit something; it's being close to people. The human contact is still the first stage for communication, so we can't lose it."

The reflections produced after the experience of appropriation and use of the social media include becoming aware of the responsibility and power of those who make the information. Actually, we have been able to observe that the participants of Seville's 15M, irrespective of their experience, developed a high sense of responsibility since being in charge of the media provides a great deal of power, and it was through these media that the voice of the movement could be transmitted; in the words of an interviewee: "I felt really small because I'm not used to these things, I'm not even a journalist or something like that, so I felt I had a lot of responsibility." As described in the introduction, the process of appropriation that people in the 15M experience, also involves thinking about the importance of developing an alternative or citizen communication. This "other communication," as defined by the interviewees themselves, arises in contrast to the communication created by the mainstream media, and envisages other values and characteristics; in the words of interviewees: "It must be spontaneous, natural, sincere, without trying to manipulate, it has to be humble. I believe it needs simple words, but clear messages."

Another idea came up from the fieldwork. It is the idea that this "other" communication, rather than being a channel for protest, must aim to develop proposals to problems that could arise in a community, as long as said proposals come from the people instead of activists in the communication committee. In fact, as the interviewees themselves explained to us, the people collaborating in the communication committee consider themselves as broadcasting tools, not as the creators of messages. This participant explained: "We wanted that people took part of what we were communicating; we wanted to be the people's tool to transmit and do things." In light of what we can read so far in our analysis, the process of appropriation and use of the digital media occurs when people value and learn to use the communicational tools to satisfy the needs and interests of their collective or community, reflecting and discussing both the means and the ends. The process of appropriation, or the relation of mediation as Martin-Barbero (1987) names it, is subjected to the culture and experiences of the people who appropriate the media to listen and recognize themselves, and it is actually from these perspectives that the protagonists develop their interpretative frames and implement "the legitimacy of the process of appropriation" (Subercaseaux 1989, 98). Lastly, as expressed in the introduction, the process of appropriation of the social media involves in itself a transformation of awareness and behavior of the people participating in the movement, that is to say, the process of empowerment, to which we will devote the third and last section of our analysis.

APPROPRIATION AND EMPOWERMENT: "IF NOBODY DECIDES FOR YOU, THEN DON'T LET ANYONE COMMUNICATE FOR YOU!"[5]

To sum up, we would like to highlight that the process of appropriation of the alternative digital media we have just described is linked to the empowerment which takes the subjects to re-create and redefine values, beliefs, and identities; to be aware of the aspects of reality they have not considered so far, to change their perception of it and act consequently. These two processes—appropriation of the media and empowerment—are neither separate nor consequential, but emerge and develop along with the experience of protest. From the first stages of appropriation, the reasons and the method, to the identification with the media and the reflection on the social media produced throughout the experience of protest that we have described above, people become empowered, transforming themselves into political subjects.

Empowerment can be both individual and collective and consists of the acquisition of power, understood not as "power over someone," but as "power to do something" as a potentiality (Dallago 2006), which is also identified as an emotional benefit in the participation in the movement (Wood 2001). As Dallago writes, "the concept of empowerment highlights, in the political arena, the narrow interdependence between individual change and social change" (2006, 11), which in our proposal of work is manifested in the cultural change after the appropriation of the social media, confirming once again the suitability of choosing a biographical approach and considering the subject's point of view, since the interdependence between the macro and micro dimensions can be observed when looking at the subject rather than looking at the structure only.

In our case of study, "the power to do" takes place in the need for creating "other" media and "another" communication, and arises as the response to a whole lot of lessons and moral shocks (Jasper 1998 and 2006) that emerge from the awareness of the collusion of the mainstream media with the political and economic powers. An outstanding example is the outrage caused by the manipulation of the news regarding the movement, which becomes one of the causes to lose legitimacy and trust in the mainstream media, as an interviewee asserts: "My vision of the mainstream media was that they said a lot of lies, everything was a lie, everything was manipulated, the television above all."

The citizens, aware of the bias of these media, then start to put forward a narrative of the reality of their own, reducing the communication of the media oligarchy's monopoly and creating their own media, as expressed by an interviewee: "The communication is totally stuck together by the structures of power and we have to take the media and communicate from the local and small sphere." The need to avail of the movement's own channels to communicate demonstrates that the appropriation of the social media is a process that is accompanied by the empowerment of people and awareness of the possibility and necessity of having a citizen communication. Thus creating their own movements, citizens do not limit to counterattack the lies of the mainstream media providing "true" information, but construct their own channel of communication with the purpose of broadcasting their ideas and values, creating new horizontal and open spaces from which the voice of the people below can be expressed; in the words of an interviewee: "The question is not confronting the system but creating and generating a system of our own . . . and I think this can only be done from the local and small spaces."

Finally, the citizens that participate in the movement experience a transformation of consciousness and behavior (Piven and Cloward 1977) that takes place in the many practices that arise from the movement. In each camp, neighborhood assembly, and so on, new groups, associations, occupations, solutions to daily problems such as access to housing or self-production of food in the city have been created. Meanwhile, in terms of access and transmission of information we have been able to confirm there were plenty of changes, as a member of the communication committee has expressed:

> From the past year and a half, in the streets it is becoming more common to talk about the alternative media; more projects of independent journalists have arisen as well as groups of people who have published on a website, magazine or Facebook wall or have made their own newspaper or radio station. I see that what is just arising, and I wouldn't say it's for us, has been a little flame that has influenced millions of factors and not a particular one. One of the issues of this revolution is that there is a high cultural level of the people who are participating. And this gives you a different perspective of how to do things.

This revolution of "how to do things" is what all the Spanish *indignados* are doing, and with this our short analysis comes to an end. We have presented how the process of appropriation of the social media can become an experience that changes people, creating political subjects better informed, more critical, less alone, more aware of their potentiality of social change.

CONCLUSION

Throughout the chapter, we have shown how, in the frame of the Spanish *indignados*' experience, the process of appropriation of the social media has been accompanied by the participants' empowerment. The approach from below, centered on the experience of the subjects has allowed us to understand how the changes are produced from the media experiences and how other changes involving the entire collectivity are produced, as people change by changing the world, as Raúl Zibechi (2007) has asserted.

Starting with the analysis of why and how the *indignados* have appropriated the social media we have been able to demonstrate that the process of appropriation and use of the media is the result of collective abilities and making something common to others. The Do It Youself practice has also taken the protagonists to a process of identification with the medium of communication, transforming the social media, as occurred with the construction of the *Toma la plaza* portal. The participants are also involved in the continuation of the actions that were already happening in the streets and in the extension of the collective imagery of the camps, which turned into laboratories of experimentation of political practices.

The major effort of the protagonists, through an action-identification-reflection process, has been positioning the social subject in the middle of the experience, surrounded by the communication tools, thus transforming the new technologies in options rather than the necessary circumstance that allows communication. Hence, with regard to the *indignados*' movement, the social media have become broadcasting channels of ideas and values exceeding the personal dimension of the protagonists and engaging the social and collective dimension.

According to what has been presented in this chapter, we can assert that one of the most interesting aspects of 15M is that in the practice of appropriation and the process of empowerment linked to the former, the limits of media spaces—the social media—and of the protest—the camps—have moved further and therefore converge in everyday spaces. The new media projects arisen in the past years, as well as the new networks of affinity groups that have been created from the experience of protest, show that participation in the movement, regardless of the attainment of the movement's claims and objectives, transforms people into political subjects who later take action to change their everyday nature using the learning they acquired during the protest.

NOTES

1. In relation to European experiences, see the results of the DEMOS project, cfr. Democracy in Europe and the Mobilization of Society led by Donatella Della Porta
2. Or as defined by Scott (1990) "occasional insurgencies" moments of explosion that many times analysts cannot explain, and it seems they emerged from nothing, while they are nourished by this hidden transcript that characterizes infrapolitics.
3. The mortgage law in Spain states that even if the occupiers leave the property because the mortgage cannot be paid, the debt to the relevant financial institution must still be paid off. Data shows that hundreds of evictions take place every day, that is, families are forced to leave their homes while their debt to the bank remains. This has had the direct impact of increasing the number of suicides, and the creation of a movement for dignified housing, which in Andalusia has taken the form of "corralas"—unused buildings that belong to banks and other financial institutions that are occupied by evicted families. For more information, cf. Stefox, 2013.
4. Demonstrations in which a group of protestors go to the homes or workplaces of those who they want to condemn and humiliate publicly.
5. "If nobody decides for you, then don't let anyone communicate for you!" This is the name of the communication protocol drawn up by the communication committee of the 15M group in Seville, a self-produced manual that forms, according to the document itself, a "tool for empowering communication and technological sovereignty."

REFERENCES

Coller, Xavier. 2005. *Estudio de Caso*. Madrid: Cuadernos CIS.

Dallago, Laura. 2006. *Che cos'é l'empowerment*. Rome: Carocci Editore.

De Certeau, Michel. 2000. *La invención de lo cotidiano. Artes de hacer*. Mexico: Universidad Iberoamericana.

Della Porta Donatella. 2013. *Democracy in Europe and the Mobilization of Society* (DEMOS, n.d.), http://demos.iue.it.

Della Porta, Donatella and M. Diani. 1997. *I movimenti sociali*. Rome: La Nuova Italia Scientifica.

Della Porta, Donatella 2010. *L'intervista qualitative*. Bari: Editori Laterza.

Flick, Uwe. 2000 "Episodic Interviewing." *Qualitative Researching with Text, Image and Sound: A Practical Handbook*, ed. Martin Bauer and George Gaskell, 75–92. London: Sage.

Flick, Uwe. 2004. *Introducción a la investigación cualitativa*. Madrid and A Coruña: Ediciones Morada and Fundación Paideia Galiza.

Freire, Paulo. 1970. *Pedagogía del Oprimido*. Mexico: Siglo XXI.

Freire, Paulo. 1980. *La educación como práctica de la libertad*. Mexico: Siglo XXI.

Hamelink, Cees. 1999. "Language and the right to comunicate." *Media Development*, XLVI (4).

Jasper, J. M. 1998. "The Emotions of Protest: Affective and Reactive Emotions in and around Social Movements." *Sociological Forum* 13 (3), 397–424.

Jasper, J. M. 2006. "Emotion and Motivation." *Oxford Handbook of Contextual Political Studies*, ed. Robert Goodin and Charles Tilly, 157–171. Oxford: Oxford University Press.

López Martín, Sara and G. Roig Domínguez. 2006. "Del tam-tam al doble click. Una historia conceptual de la contrainformación." *Ciberactivismo. Sobre usos políticos y sociales de la red* , ROE. Barcelona: Virus.

Marí Sáez, Victor. 2004. *La red es de todos. Cuando los Movimientos Sociales de apropian de la red*. Madrid: Editorial Popular.

Martín-Barbero, Jesús. 1987. *De los medios a las mediaciones. Comunicación, cultura y hegemonía*. Barcelona: Editorial Gustavo Gili.

McKay, George. 1998. *DIY Culture: Party & Protest in Nineties Britain*. London: Verso Books.

Pasquinelli, Matteo. 2002. *Media Activism. Strategie e pratiche della comunicazione indipendente*. Rome: Derive & Approdi.

Piven, Frances F. and R. A. Cloward. 1977. *Poor People's Movements. Why They Succeed, How They Fail*. New York: Pantheon Books.

Proulx, Serge. 2004. *La Révolution Internet en question*. Montréal: Editions Québec Amérique.

ROE (Reunión de Ovejas Electrónicas). 2006. *Ciberactivismo. Sobre usos políticos y sociales de la Red*. Barcelona: Virus editorial.

Romero, Maria Angel, and M. Dalton. 2012. *Para que NO se olviden. Mujeres en el movimiento popular. Oaxaca 2006*. Oaxaca, Mexico: Secretaria de Cultura y Artes.

Sanz Hernández, M. Alexia. 2000. *Ojos Negros. La memoria de un pueblo*. Teruel, Spain: Instituto de Estudios Turolenses and Ayuntamiento de Ojos Negros.

Scott, James C. 1990. *Domination and the Arts of Resistance. Hidden Transcript* (New Haven and London: Yale University Press).

Sierra, Francisco. 2012. "15M, Occupy Wall Street, Ocupa Brasil." Paper presented at 2nd *Colóquio Brasil Menor*, Brasil Vivo, São Paulo, Brasil.

Spencer, Amy. 2008. *DIY: The Rise of Lo-fi Culture*. New York: Marion Boyars.

Stefox, Dave. 2013. "How the corrala movement is occupying Spain," *The Guardian*, March 22, http://www.guardian.co.uk/world/2013/mar/04/corrala-movement-occupying-spain.

Subercaseaux, Bernardo. 1989. "Reproducción y Apropiación: Dos modelos para enfocar el dialogo intercultural." *Revista Diálogos de la Comunicación* 23, 57–62.
Wood, Elisabeth J. 2001. "The Emotional Benefits of Insurgency in El Salvador," in *Passionate Politics: Emotions in Social Movements*, ed. Jeff Goodwin, James M. Jasper, and Francesca Polletta, 267–281. Chicago: University of Chicago Press.

Chapter 3

Occupy Nigeria Movement, Organized Labor Unions, and Oil-Subsidy Struggle

An Analysis of Processes in Media(ted) "Revolution" and its Demise

Kudus Oluwatoyin Adebayo

The wave of protests of the last couple of years in different parts of the world, and the changes that accompanied it in some African, Asian, and Middle Eastern countries, have once again brought to the forefront of academic discourse the subject of the revolutionary potentials of social movements in the age of new/social media. The new order of political consciousness and socioeconomic awareness engendered by the social movements of the late twentieth century, and which crystallized in the forceful and large-scale explosion of the Arab Spring, Occupy, and *Idignados* movements in this century have fascinated political scientists, sociologists, and scholars in communications and development studies. Online depositories of scholarly commentaries, analyses, databases, and academic journals have been created to capture what is occurring and to document events as they unfold. Among other issues, scholars have dedicated a great deal of effort towards describing the technological, social, political, economic, and even cultural processes involved in these movements, providing , for why some succeeded (or failed) while, at the same time, constructing and conjecturing about their potentials in the pursuit of freedom, democracy and good governance.

To date, a crucial note of caution that resonates across academic discussion of the new social movements, which is taken as a point of departure in this present paper, relates to the need for contextual understanding of these events, even as each and every one of them appears to be linked and similar in motivation, form, and method. For instance, while acknowledging the similarity in the goals that Occupy movements articulate in their various sociopolitical and geographic settings, Kennedy (2011) emphasizes that each of the struggles has its own powerful and peculiar history. In raising the question of the possibility of the so-called African Spring in the wake of citizens' revolutions in the Arab world, Ford (2012) cautions against the temptation of expecting similar conditions of unemployment, poverty, official corruption' and other social ills to lead to uprising in sub-Saharan Africa. Even within the "Arab Spring" collective, Ali Mazrui (Mazrui and Tanoukhi 2011) was sure to emphasize the internal dynamics of the North African region, pointing to the similarities and relationship between revolutionary movements in Tunisia and Egypt on the one hand and in Libya on the other hand. This is in spite of the fact that these countries have a high degree of similarities of history and culture and shared deployment of social media tools while the revolts lasted.

In examining the Occupy protest as experienced in Nigeria, this article seeks to contribute to the growing body of knowledge on dynamics of twenty-first-century social movement. The paper aims to provide a context-based description and analysis of a social protest that was inspired and influenced in important ways by global events and culture but whose processes was largely shaped by a history and geopolitics that were internal to the Nigerian society. Using the January 2012 oil-subsidy struggle as case study, two general objectives are pursued in the article: first, a descriptive objective focusing on how organizers of Occupy Nigeria Movement (ONM) stimulated the revolutionary capacity of an apathetic population, mobilized them to action, and coordinated on-street activities in Nigeria and across cities around the world, mostly through the use of new media forms; second, an analytical objective aimed at dissecting how the power structures and privileges of organized-labor unions (OLUs) enhanced and later undermined the most massive social uprising in the history of modern Nigeria.

The paper draws from primary and secondary sources, including relevant literatures, news reports, social media sites, protest documentaries, online and offline conversations, and interviews with Occupy Nigeria (ON) mobilizers and labor union officials. The remainder of the paper is organized as follows: immediately after this introduction, the national context of Nigeria is presented in section 2; section 3 considers how the rise of Internet and new media technologies contributed to the resurgence of public sphere in the country; in section 4, the discussion is grounded theoretically by dialogically exploring how social actions that are mobilized in the new social media-mediated public sphere were still dependent on the vagaries of interests and power, both of which play out primarily offline rather than on the Internet; sections 5 and 6 are respectively devoted to the descriptive and analytical objectives set out earlier, while the final part concludes the paper.

THE NATIONAL CONTEXT OF NIGERIA

Nigeria is a West African nation of over 170 million people (Internet World Stats 2013), the largest, most populous black nation in the world. An ethnically diverse society, Nigeria is rich in both human and natural resources, occupying an area of 923,768 sq. km. with massive landscapes that can support all manners of economic activities, especially agriculture. Since gaining independence from the British colonialists on October 1, 1960, the country has been a ground for experimenting different political forms, although close to three decades of self-governance was spent under military rule, starting from 1966. Before oil became the mainstay of national economy, the people were actively involved in agriculture, exporting crops such as cocoa, kola nut, oil palm, cotton, and so on. When exploration for oil began, agriculture was gradually relegated to the background, and much of the food consumed in the country today is imported.

In spite of huge revenues from oil, chronic poverty is widespread among the population. For instance, close to 70% of Nigerians were estimated to be living below the poverty line of $1/day in 1999, up from 27% in less than two decades (Nigeria 2004). A publication by the National Bureau of Statistics (NBS) also showed that Nigerians were generally poor, regardless of the measure employed, with about two-thirds of her citizens living in poverty (NBS 2012). While the country consistently showed signs of quantitative growth, with analysts believing in the eventual transformation of this success into actual human capital development, the challenges of corrupt and profligate public service persistently make "positive development" a mirage. As a result, needed foreign direct investment, which holds promises of both formal and informal employment for the population, has not been forthcoming. Since 1999, the country had enjoyed some fourteen years of political stability in the form of democratic governance under a single ruling party. Yet, a bulk of Nigeria's youthful population are unemployed.

Today, the oil sector provides about 95% of foreign exchange earnings, and close to 80% of budgetary revenues (Index Mundi 2013). Unfortunately, the wealth has benefited the political class more than the people. Nigeria's economy of oil gave rise to a rentier state in which the government became more and more dependent on oil revenues, and developed a crass sense of irresponsibility and unaccountability (Gboyega, Søreide, Le, and Shukla 2011). This natural resource is also a major driver of corruption in politics and in publicly owned establishments (Nwokeji 2007). To a great extent, the wealth derived from oil has fueled poverty among the population, and painfully so in the creeks of Niger-Delta, an oil-rich region where the most dehumanizing realities are commonplace (Obi 2009; 2010).

Meanwhile, local refineries have since failed to even produce enough refined oil for local consumption. The paradox of this condition is that citizens now depend on the vagaries of international markets to consume what they produce at affordable rates. Fuel subsidies were first introduced in the 1980s to cushion the effect of the high cost of refined crude in the global market. It was supposed to be a temporary policy but over the years, subsidies on petroleum products have become permanent, and an avenue to perpetrate a form of institutional corruption for which the oil industry is currently renowned. More significantly, however, is the fact that the subsidy on petroleum products had transformed from mere cautionary policy to a sort of social benefit; the most obvious social benefit enjoyed by the people from the government. As successive governments failed in their capacity to develop social and economic infrastructures and provide good leadership, people have grown apathetic, expecting little from political officeholders while finding respite in purchasing petroleum products at prices below international levels.

INTERNET, NEW MEDIA, AND NIGERIA'S RESURGENT PUBLIC SPHERE

The context of pervasive social and economic decline, and the attendant misalignment within the political realm, succeeded in destroying the public sphere of the Nigerian society. Over a period of five decades, the once vibrant public sphere was haunted by numerous corporal ghosts, but those of militarism and underdevelopment were most impactful. On the one hand, the public space that accommodated public discourse and that led to nationalist mobilization for freedom and national independence was effectively neutralised as governance became the domain of military dictators. Although labor unions and civil society organizations challenged military rulers and protested against "anti-people" policies, Aiyede (2003) notes that most of them either were too alienated from the state or tended to be too far removed from the grassroots where active citizenship ought to have been promoted. For him, military governments shaped civil society more than the latter shaped political events. On the other hand, an unending problem of policy somersault or what Akanle (2013) called "development exceptionality," coupled with corruption, profligacy, and impunity among public officers, created a condition of apathy that deincentivized popular participation in political affairs.

While the return to democracy on the eve of the millennium opened up some space for public discourse to take off, the "real" transformation of the public sphere did not begin until more and more Nigerians ventured into the cyberspace. It should be noted that this process dates even further back than May 1999 when the military handed over power to civilians. In an ethnographic study by Bastian (1999) and in a recent book by Everett (2009), Nigerians in the United States were reported to have been among the few earliest callers on the cyberspace among black communities, with strong online presence as early as the 1990s. Through discussion forums, information sharing, and publication and distribution of position papers, these cybercitizens of Nigeria helped stimulate a dying public sphere and provided the pillars that would later speed up citizens' recovery from the slumber of political disinterest.

The subsidy-removal struggle of January 2012 received a great boost from the recent increase in the number of Internet users and the changes that occurred in the telecoms sector. This transformation led to the rapid spread and adoption of new/social media tools and networking sites by Nigerians at home. From 2000 to 2008, Nigeria added 11 million new Internet users, representing close to 40% of total additions in sub-Saharan Africa at the time (Aihe 2010). While that number was estimated to represent a penetration rate of only 7.4% in 2009 (SPARC 2010), the figure had increased to over 48 million Internet users in June 2012, a penetration rate of 28.4% of the population (Internet World Stats 2013). This surge was due mostly to the deregulation of the telecoms sector from 2001, which not only increased the number of people with mobile phones but also made it cheaper for those using devices with minimal Internet capability to access the cyberspace.

Other important contributory factors in opening up the public space, as a result of the Internet revolution during the post-1999 years, include innovative use of simple mobile phone technology such as short messaging service (SMS) and picture and video apps, the influx of smartphones, and the popularity of social media and networking sites among Nigerians. All of these were of course interdependent in practice. Before the protest against subsidy removal in 2012, civil society groups, as well as low-scale, grassroots advocates and mobilizers, have explored the powers of SMS and social media sites during the general elections of 2011. Through mobile phones, Nigerian youths or "the thumb tribes" as Taiwo (2010) describes them, texted election results to friends as they were announced at the wards, and broadcasted same through the messaging platform of smartphones like Blackberry. Most of the results ended up on new media platforms such blog sites (Word Press and Blogger), social media sites (like Facebook and Twitter), and news-based sites like Osun Defender, Sahara Reporters, Nairaland, and so on.

Indeed, the ease with which citizens were able to access and share large volumes of information over the course of the election period, as well as the feeling of "we-made-it-happen" that became dominant in post-election discourse of the 2011 elections, were enough to revive the lethargic public sphere. Although politics of corruption and impunity remained rampant, and poverty was still on the rise, the awareness that political participation was yet a possibility—and perhaps cheaper, easier, and less risky—reverberated in the consciousness of millions, especially among the younger generation who have become increasingly restless about the status quo.

THEORETICAL STATEMENT: THE NEW PUBLIC SPHERE AND THE PUBLICS OF INTEREST

The extent to which the Internet may be meaningfully described as a potential public sphere has been a subject of academic debate. Relying on Jürgen Habermas's influential book titled *The Structural Transformation of the Public Sphere* (1962), scholars engaged in the discussion with aims to identify and highlight the conditions and the set of attributes that can transform the cyberspace into a sphere of communicative discourse and deliberation. Properly defined, public sphere constitutes a space of action involving varieties of public sites where people come to deliberate on common concerns, evaluate competing proposals for change, and eventually arrive at a public opinion (Gripsrud, Moe, Molander, and Murdock 2010). Implicated in the idea of public sphere, according to Gripsrud, et al. (2010), is a notion of "public" that involves concrete physical space, as well as commonly shared and accessible informational and cultural resources; issues that are legitimated as governance concerns because of their intersection with interest common to all; a social category for collectives of citizens; and an aggregate of individual view which emerges as public opinion. As an action-oriented activity, opinions that emerge in the public sphere should normally shape politics and public policy in any democracy.

Against this background, Bohman (2010) highlighted some factors that must necessarily be present before a sphere is pronounced public:

> First, a public sphere . . . must be a forum, that is, a social space in which speakers may express their views to others and who in tum respond to them and raise their own opinions and concerns . . . Second, [it] must manifest commitments to freedom and equality in the communicative interaction in the forum. [Also,] communication must address an indefinite audience. In this sense, any social exclusion undermines the existence of a public sphere (Bohman 2010, 249–50).

Scholars that are optimistic of the public sphere status of cyberspace argue that the Internet possesses these features and even more. As authors such as Di Maggio et al. (2001), Cohen (2007), Mayes (2009), and Bohman (2010) have tried to explain, optimists are generally of the opinion that features such as anonymity, inclusiveness, simplicity, relative cheapness, hyper-interactivity, speed, ease of role switching and taking by former audience, many-to-many communication and so on all position the Internet as a credible sphere for public engagement and democratic deliberations, even though the interaction does not take place in the salons or coffee houses. In Mayes' view.

> While Habermas spoke of public places such as the coffee house, he also conceived of a public space where debate occurred in which people abandoned their private interests and engaged with issues they considered to be in their collective interest. In the same way internet technology informs a sphere, a virtual space, imagined as egalitarian and theoretically accessible to all and sustained by anonymous expression (Mayes 2009, 94–95).

The dialogic character of cyberspace makes it an egalitarian space (Bohman 2010). It also appears that these reasons explain why contemporary social movements are increasingly converging around the Internet to construct and share imageries of a much freer world and also mobilize for it through the same means.

Nevertheless, some scholars warned that caution must be exercised in designating the Internet as public sphere. There are two broad interrelated strands of arguments to this call: the first set dealsng with the theoretico-conceptual convergences and divergences that underlie the "idea of public sphere" as conceived by Habermas and the structure and infrastructure of the Internet; the second set of arguments raises substantive and pragmatic questions that seek to understand the limits and the real potentials of Internet as a sphere that can inform concrete social actions, and lead to societal change. The first set of arguments was engaged by Cohen (2007), Bohman (2010), and a few other scholars, and it is beyond the scope of this article to discuss them. Of more direct relevance to this work are the substantive issues.

Since the Internet has become so attractive to social movement, and considered by many (cyber)activists and "digital utopians" to have symbolic and practical relevance for bringing about social, economic, and political changes, the new public sphere of Internet and the associated new media forms have come under scrutiny. Aside from the obvious threats to participation in the cyberspace—linked mainly to the issues of digital divide, illiteracy, surveillance, and censorship—(Banda, Mudhai, and Tettey 2009) there are concerns about the ability of cyber-activists to translate online activism, and the mass mobilization achieved through the platform, into the kind of results that reflect their own aspirations and goals. From North Africa to Europe and America, activists that mobilized for occupation and mass protests cannot confidently claim that articulated values and visions are being seriously considered in the post-protest agenda of those in power. Apart from injuries caused to the vision of cyberactivists, and the political cynicism expressed toward their mobilization and tactics of occupation, the practicability of the prefigurative world that discussants constructed via the Internet was often discursively downplayed and undermined by power brokers and institutional guardians who often accused them of too much utopia (Zuquete 2011). These power brokers are offline actors that benefit from the existing arrangement and must, as a matter of course, ensure that the status quo is defended at all cost.

Publics of interests such as the government, large businesses, organized unions, and other similarly patterned institutions continue to dominate offline affairs and structures. Also, the desire of cyberactivists to establish a free, autonomous, and "reality-altering" sphere on the Internet is still in one way or another connected to the preferences of these publics of interest. The emergence of the Islamic Brotherhood in Egypt and the pursuit of a religionist agenda that contradicts the vision of democracy that protestors articulated is a case in point. What seems to be missing in the calculation of participants of the Internet public sphere is the neglect of the fact that cyberspace is, first and foremost, part of lived space (Cohen 2007). Normally, transformational strategies to be carved ought to be based on an awareness of the connections of online space to lived space.

FUEL-SUBSIDY REMOVAL AND OCCUPY NIGERIA PROTEST: PROCESSES AND ACTORS IN MEDIA(TED) "REVOLUTION"

On the first day of January 2012, the Nigerian government, through the Petroleum Products Pricing Regulatory Agency (PPPRA), announced the removal of the subsidy on petroleum products. Prior to 2000 when the first pronouncement was made under a Fourth Republic civilian government, the subsidy on petroleum products has been removed more than eight times since 1978, mostly by military dictators. The first main hike in petroleum prices was in 1978, from 0.8k to 0.15k.[1] In the first decade after that, it went up by approximately 67% to 42k. By 1999, when the first democratic election was held since 1983,[2] petroleum was selling for ₦20, a 98% rise from the price in 1989. Prior to the recent subsidy removal, civilian government had increased the price of PMS (petroleum motor spirit) by another 69.2%, selling at ₦65 on the eve of 2012.

As such, the removal of the subsidy on fuel and other petroleum products is not new in Nigeria and to Nigerians. In fact, public discourse and official activities around subsidy removal or nonremoval intensified since structural adjustment became the economic order, used at different times as instruments of threat, resistance, or political legitimation rather than merely as tools for managing government deficit.[3] The normal response of Nigerians to these changes was to resist by protesting or by embarking on strikes, often through the involvement and leadership of OLUs, rights movements, civil society, and advocacy groups. However, until the 2012 protest, only on one occasion did an anti-fuel subsidy removal campaign managed to last more than eight days. Previous petroleum-related struggles also suffered from poor coordination and were not as transethnic, interclassed, intergenerational, and transnational as the uprising that was triggered by the subsidy removal of January 1, 2012. A number of commentators believe that the protest of 2012 was the largest and most sustained short-term protest in any sub-Saharan African country in a long while (Jacobs 2013). The rest of this section describes the protest, its form, the actors involved, their methods, and the role played by social and new media tools over the course of the campaign.

OCCUPY NIGERIA MOBILIZERS AND ORGANIZED LABOR UNIONS IN THE SUBSIDY REMOVAL PROTEST

At the heart of the social protest that grounded national economic activities for more than two weeks in Nigeria, from January 2 to 16, were feelings of discontent with governance failures, institutional corruption, and the insensitivity of the political class to the poverty condition of the people. Expressing a motivation that captures the feeling of most young Nigerians, Japheth Omojuwa, the initiator of #OccupyNigeria hashtag on Twitter, stated that the protest was natural: "The time had come for the Nigerian government to realise and understand the country is not about 'them' alone but about 'us.'"

Although many groups participated in the struggle, ON mobilizers and OLUs were most prominent. Among the first group were individual online and offline activists; civil society organizations working in the areas of governance, corruption, political integrity and so on; public figures, businesspeople, actors, musicians, and so on. What most of them had in common was that they subscribed to the idea of "occupy" while articulating their respective views on the rationale for the protest. The other group consists mainly of structured workers' unions across different sectors of the Nigerian economy, represented by large umbrella bodies like the Nigeria Labor Congress (NLC) and Trade Union Congress (TUC). NLC and TUC are the largest labor unions in Nigeria, and their members account for the highest number of workers in the formal economy. But in order to extend their outreach and influence in the larger society, both unions formed the Labor and Civil Society Coalition (LASCO), with the "radically ideologized" Joint Action Front (JAF) as its third member. JAF is made up of hundreds of small- to medium-scale civil society organizations across the country. As will be shown in subsequent sub-sections, this distinction is important for comprehending the processes involved in Nigeria's subsidy protest of January 2012.

OCCUPY NIGERIA MOVEMENT VERSUS LABOR COALITION: CONTESTING THE PROTEST SPACE?

For the most part, individuals and groups that were not on the side of government worked together. However, the emergence of young activists that identified with the philosophy of global Occupy movement appeared to have posed some threats to the established order of protestation in Nigeria. Members of OLU tried frantically to dissociate themselves from the Occupy-movement, in a bid to mark territory and proclaim dominance over the Nigerian protest space. For example, in-depth analysis of JAF's blog at jointactionfront.blogspot.com showed that the group was very much interested in ensuring that people understand who controls or "owns" the protest. In describing the role played by the organization, a high-ranking executive was very direct and emphatic: that JAF was the initiator, organizer (with its labor partners—NLC and TUC)—mobilizer, and determinant of the nationwide protests. When probed further to describe what connections exist between JAF and ONM, he elaborates thus:

> The issue to note is to have a conceptual clarification of the 2012 January uprising beyond the international profiling of what is called "occupy movement." . . . [W]hatever is meant by 'occupy' should be contextually defined because the term "occupy" is an understatement of our continuous struggle of resistance for the socio-economic transformation of Nigeria.
>
> Nigeria's realities are completely different from the western media profiling of the January uprising. *We maintain that the 2012 January uprising has a clear leadership and shouldn't be reduced to an abstract categorisation such as 'occupy.'*
>
> Our protests - JAF and its Labor allies have always been Mass. *There was nothing different in the mass character of the Jan (sic) 2012 uprising than previous massive protests other than the international profiling and public expectations that it would have gone in the direction of so-called Arab Spring.*
>
> . . . Nonetheless, we took note of massive participation in the January 2012 uprising by such groups as Save Nigeria Group (SNG) and others who categorised themselves as 'occupy movement' but it is not our business in JAF to determine what role any organisation plays in protest, what is primary to us is working with all groups that are agreed to our goals [emphasis added] (Executive Secretary/JAF/Online Interview/September 2013).

The new generation activists that promoted ONM, and that mobilized for and coordinated protest efforts through social media sites, announced their entry into the protest space of Nigeria by targeting younger citizens. ON organizers engaged this group and encouraged them to not only discuss issues but to also "occupy." This approach was significant, at least from the viewpoint of the fact that it permitted younger citizens to construct, through sustained online discourse, narratives that rationalized and justified their involvement. Organized labor and traditional civil society groups have failed to provide this for a long time (Democratic Socialist Movement 2002). The facticity of shared sentiment and comparability of lived experience in the Nigerian state enhanced trust-building between ON mobilizers and youths all over the country. To an extent, most of the highly educated, upper middle-class citizens who had developed great cynicism towards organized unions were attracted by the strength of unity displayed by the youths.

The fact that ON mobilizers were as active and as crucial to the protest as "old" labor unions was demonstrated on the streets of major cities across the country. Contrary to the approach of the past, ON mobilizers focused less on moving from one point of the city to another. Following the method of the global Occupy movement, symbolic public places were occupied, including parks, expressways, and public office buildings. Occupy protestors also carved out temporary autonomous zones in some cities, particularly in Kano, Ibadan, and Abuja. Protestors held on to these spaces and used them as base stations for coordinating protest efforts. Government responded to these zones by imposing curfew (in Ibadan) and with direct violent attacks (in Abuja). At Eagle Square, a massive space at the centre of Abuja, hired thugs attacked and assaulted protestors in the middle of the night, all in the presence of the police.[4]

ON mobilizers were quicker responders, and their approach to protest is simple, yet effective. The new entrants acted more freely and spontaneously, unlike OLUs who were constrained by size and structure, as well as their position in relation to the law. Why OLUs were determined to appropriate the protest space is still not clear; what is worthy of note in this article is that such dynamics manifested in the subsidy protest of January 2012.

ROLE OF NEW/SOCIAL MEDIA

Social media played an important role in the subsidy struggle of January 2012. However, it was not deployed by all the participants. For those that used social media platforms to enhance their activities, it was not employed equally, and the level of importance attached to it differed.[5] Whereas OLUs used social media as "only a part of the tools . . . to enhance information dissemination to the general public," ON mobilizers were distinguished by the extent to which they exploited Internet sites such Twitter and Facebook. In addition to dissemination of up-to-the-minute information across multiple platforms, easy-to-know hashtags (e.g., #OccupyNigeria, #eyesANDears) were created to facilitate discussion among protestors on the streets of various cities. In pre-protest periods, group e-mail was in constant use and on different occasions, virtual meetings involving many people across multiple centers were also held. As one participant explained, social media: "helped to amplify our voices within and outside the country. Some 32 cities around the world joined in the movement and only social media could have helped with such a spread especially considering the time limits" (Omojuwa 2013).

Before January 1, 2012, when the announcement was made, government was supposed to be "consulting" with relevant stakeholders on the need to remove the subsidy. What the government had agreed to do was to hold off on any decision until April 2012, and many Nigerians had been led to believe it. As such, the decision by government to remove the petroleum subsidy on New Year's Day caused an upset within ON camp. By also making that call at a period when many Nigerians have travelled to their hometowns and villages, government also gained the advantage of truncating the access of anti-subsidy removal mobilizers to their most important asset: the people.

ON activists had no plan to begin "occupation" in January, but when that became their only opportunity to resist, new/social media tools gave them the leverage they need to mobilize, plan, and coordinate across spaces. On January 2, ON activists were able to mobilize hundreds of people to appear at Eagle Square, Abuja. Although their intention was to get people together to sign petitions physically and compile volunteers lists, government saw the threat posed by their efficiency and responded by deploying security forces to teargas, harass, arrest, and disperse them. At this time, no concrete response to the subsidy removal had been announced by the largest organized labor unions in the country (NLC and TUC). Apart from the JAF that had organized some rallies at the southwest city of Ibadan, NLC and TUC did not mobilize until January 9, 2012. The swift response of ON mobilizers aligned closely with the flexibility that most Occupy protests were known for. Indeed, the ability to be spontaneous and mobilize under serious constraints had been observed to be one of the most important benefits that social media adds to the effectiveness of contemporary social movements (Petras 2011; Wilson 2012).

In addition, personal blog pages of ON organizers were used to engage members of the public who were yet to decide whether to join the protest or not. They educated would-be protestors on legal-rational rules guiding protest, as stated in the constitution and other codes, and advised them on ways to manage affairs at the home front while the protest lasted. Again, rather than wait for TV stations or newspaper reporters to come to them, ON activists packaged their own pictures, recorded short videos, and uploaded them on different websites for easy and fast consumption. Links to these materials were shared on social networking sites, microblog pages, and through instant messaging apps on smart phones, particularly Blackberry.

The important role played by Internet, as well as the associated new and social media tools, is also deducible from the transnational character of the protest. In London, New York, Washington, Brussels, and Toronto and in Switzerland, Australia, South Africa, and elsewhere, Nigerian diasporians protested economic policy at home and solidarized with their countrymen and country women in transnational space. With ease of access to information about events as they developed back home, Nigerian nationals abroad joined forces with fellow African diasporans and Occupy apologists in various cities to denounce what they called government's insensitivity and high-handedness, while also rallying international support for protestors. In Washington DC, public intellectuals such as Emira Wood of the Institute of Public Policy and Sameer Dossani of ActionAid marched across Macpherson Square to the White House before stopping at the World Bank to deliver the "We are 99%" speech. Demonstrators in New York also occupied public spaces at the United Nations. Transnational online news media like Sahara Reporters helped to broadcast up-to-the minute reports of happenings in Nigeria and in global cities elsewhere. Although many online news media covered the protest, Jacobs (2013) observes that Sahara Reporters had a much larger impact. By uploading pictures, articles, and in-depth analysis of the petroleum industry by experts on its website, Sahara Reporters provided a platform through which Nigerians from all over the world interfaced with events at home and transnationally voiced solidarity with their fellows.

POST-PROTEST DISCOURSE AND THE "BETRAYAL RHETORIC": WHEN ONLINE ACTIVISM MEETS OFFLINE POLITICS

The protest against subsidy removal started the next day after the announcement was made, but its impact and seriousness was not really felt by government until OLUs embarked on strike action on January 9, 2012. The emergence of NLC and TUC on the protest space gave the mass movement the boost it required to demonstrate people's resolve to resist what they tagged "antipeople policy." Their involvement also forced the House of Representatives out of recess to hold an emergency meeting on Sunday, January 8. The admiration of the strength of organized union as a powerful force transcended government circle to include ON mobilizers who admitted that NLC/TUC provided leadership and transformed the official attitude towards a protest that only a few political elites hitherto considered threatening at the initial stages. However, on January 15 when the strike was called off by NLC and TUC after a closed-door negotiation with the government, many had started to question the value of organized unions.

After the protest, "betrayal rhetoric" dominated public discourse because of the anti-climactic fashion with which the protest ended. Particularly, the role played by the organized labor union in its demise had been a subject of public debate and journalistic speculation. Because the protest was going so well on the streets, millions of Nigerians were hopeful that a new socio-political and economic order was possible. Most participants on the streets believed they were winning and assumed that government would have no choice than to reverse its decision after a few more days of strike and protest. Thus, the suspension of strike was, from the viewpoint of the people, needless at the time it happened. Describing how labor impacted the protest, one ON mobilizer stated that "They impacted it positively earlier on by ensuring workers joined the protests without the risks of being sacked and then they had an even more eventful impact by selling short to the government." When probed to get his view on whether organized unions hijacked and sabotaged the protest, he responded as follows: "Hijacked? No. Sabotaged? Absolutely. It took the call-off of the nationwide strike earlier called by Organized Labor to quench the spread of the protests. Labor cost the protest potential victories by yielding ground too easily and too early." Interestingly, JAF, a civil society coalition group that was in partnership with NLC and TUC, also issued a public statement formally dissociating itself from the decision. Part of the press statement reads

> The Joint Action Front (JAF)—the pro-labor civil society partner of the Nigeria Labor Congress (NLC) and Trade Union Congress (TUC) in the Labor and Civil Society Coalition (LASCO) and its allies . . . dissociate self from the declaration by the President of the NLC and TUC for the suspension of the Strike/Mass Action that commenced on January 9th as a joint initiative of the NLC, TUC and JAF . . . .[We] strongly deplore the suspension of the Strike/Mass Action as endorsed by Presidents of NLC and rue as a betrayal of this legitimate demand by Nigerians that fuel price must revert to N65 as a condition for negotiation (Joint Action Front 2012).

Due to how NLC/TUC had handled subsidy matters in the past, the call-off did not surprise many people. In some quarters, such move was even expected. What most did not however anticipate was the impact the decision would have on the struggle in its entirety. The nationwide strike action that was called by organized labor unions did not start until January 9, more than seven days after ON mobilizers and civil society groups like Enough is Enough and Save Nigeria Group had stormed the streets with thousands of marchers and occupiers. In other words, organized labor unions (that is NLC and TUC) were not the initiators of the protest. Yet, at the point when they suspended the strike, it created the impression that the mass resistance had been suspended by all other actors, and effectively gave the government a free hand to declare those found on the streets after January 15 as peace disturbers and law breakers. To a large extent, the agreement brokered between government and organized labor unions was also what gave the government the audacity to deploy soldiers to occupied sites and what permitted the authorities to apply open force on protestors from January 16.

In post-protest discourse, most analysis had bordered on finding out "what" motivated OLUs to accept government's proposal. To this, the possibility that government bribed union leaders was most speculated on, although reports of arm-twisting, national security concerns, and politicization of protest by opposition parties were not uncommon. In justifying the agreement, for instance, a labor leader said that the demand that President Goodluck Jonathan should step down was interpreted by government as a national security issue, stressing that "it was a misplaced assumption of concrete realities in Nigeria by local and international media in correlating the Ojota-Lagos mass gathering with the political power contestation in Tahir square Egypt. The leadership (NLC-TUC-JAF) of the protest movement was never agreed to unseat the neo-liberal puppet government in Nigeria. The general focus of the Protest was TOTAL REVERSAL of the price hike of petrol." So when the interpretation was tabled during negotiation, unions had no other choice than to settle so as to protect the people from violent repression.

While the importance of the "what" question is not denied, a more pertinent question from the view of the author would have been to understand "why" it was possible for OLUs to end a protest they never started.

Understanding the "why" question surely depends on several issues that may not be immediately obvious. Some of the issues may even need a much fuller, lengthy, and elaborate analysis, all of which the space available to us in the paper is not enough to accommodate. Nonetheless, I wish to point out quickly that a necessary part of the answers is the fact of the strong hold that organized labor unions had on the offline world. For the most part, online campaigning against subsidy removal and sustained construction of alternative courses of action through an Internet-mediated public sphere helped ON mobilizers to simulate the revolutionary capacity of the people. It also helped in no small measure to organize and eventually *get people on the street*. However, keeping them on the street was a task that depended more on offline activism and real-life politics.

The recognition of organized labor unions by law and the bureaucratized structure with which their activities were conducted positioned them far above Occupy mobilizers. As institutionalized bodies, they act and speak as though they represent every working person in the country, even though they had no jurisdiction over millions of workers in informal employment and countless other working persons in contract employ. Apparently, the government is familiar with this "representational impressionism" and subliminal acceptance of same by majority of the population. As such, influencing or coaxing them into accepting a deal and causing them to suspend the strike seemed the surest way of making the mass feel that protest had ended, even though the struggle could have continued in spite of the strike call-off.

At a more pragmatic level, some of the ON activists that I conversed with mentioned that their lack of representation at the negotiating table was a critical factor in the anticlimactic end to the protest. The offline politics of space closure to noninstitutionalized "bodies" that were produced via communicative discourse on the Internet public sphere was apparently meant to limit the influence of online influencers to the cyberspace. While cyber activists and unions depended on each other to articulate a vision of dissention that will force the government to take positive action, the failure of the former to perceive the counterrevolutionary potentials of the latter prove they were too inexperienced to battle offline as much as they did online. Also, the extension of social capital, created on the cyberspace by online actors, to offline organized unions without cautiousness also shows the naiveté running through the system of occupiers.

CONCLUSION

When the fuel subsidy removal struggle was brought to an end in the most anti-climactic fashion by organized labor unions, participants have been taking time to reflect on what occurred and why things turned out the way they did. Of even greater concern was how to respond to questions about how to properly place the outcome of the protest in terms of success or failure. For a good number of ON collaborators, the protest of January 2012 cannot be boxed in one to the exclusion of the other; focusing on outcome alone, without looking at the processes involved and the symbolic relevance of popular uprising, underplays the significance of the collective action. As one mobilizer observed,

> A point was made by the protests . . . that essentially showed the Nigerian government the capabilities of the Nigerian people. Many had said such would never happen in Nigeria but it did happen. That has been a disincentive for some other sinister ideas the government would have unleashed on the people even after that period but for the understanding that they could start a fresh movement that may be hard to stop considering the protesters would have learned not to depend on labor (ON Mobilizer/Online Interview/September 2012).

From the viewpoint of ON activists, the act of "statement making" is in itself revolutionary, at least in the context of Nigeria where governance failures and widespread impunity among public officeholders had repressed the civil society to perpetual passivity. Occupiers comprehend their success within a broader frame of resurgence of alternative *cybersphere* and the possibilities it holds for influencing or shaping politics in the traditional public sphere. While admitting that failures resulted from the spontaneity of the protest, which affected their level of organization, occupiers see that the rebirth of civil society and public participation were significant and worth celebrating.

In conclusion, it must be mentioned that there are aspects of the processes teased out in the article that need to be subjected to more in-depth analysis in future research. For example, more detailed study is required to fully understand why organized unions would be interested in appropriating the protest space for themselves. The tension that will result from this would surely impact the relationship between established mobilizers and the newly emerging cyberactivists, and impact how much they will be able to collaborate in the future. Other research questions may be asked: How can the friction between online activists and organized labor unions be straightened out? What factors will contribute to success? And as new media tools continue to penetrate the population, and opening up the cyberspace to people through social media and networking sites, what can we expect the Nigerian protest space to look like in the future? What possibilities exist in this space for promoting public welfare and good governance?

NOTES

1. Currency units of Nigeria are naira (₦) and kobo (k). 100k equals ₦1.

2. Note that an election was held in June 1993 but was annulled by the then military ruler Ibrahim Babangida.

3. As instrument of threat, General Ibrahim Babangida used "subsidy removal talk" to chide interborder smugglers of petroleum products and employed it frequently in growth and national development discourse (Nwosu, 1996). His successor, General Sanni Abacha, stalled many times in deciding whether to jerk up petroleum prices or not and even reinstated a substantial portion of the domestic fuel subsidy within a few days of taking over power as a show of resistance toward the neoliberal policies of the West (Lewis, 1994). During the Fourth Republic, an upward review of subsidy by Olusegun Obasanjo a few days before leaving office in 2007 was reversed by his replacement, Umar Musa Yar'Adua, as a strategy to legitimize an election that was widely claimed to have been fraudulent.

4. See "Attack on #occupyNigeria Abuja Protesters" at http://www.youtube.com/watch?v=MVT7l0C2Mgw. Last accessed on September 13, 2013. "On the 4th Day of #OccupyNigeria, the Police came to Attack Innocents" at http://azeenarh.wordpress.com/2012/01/09/on-the-3rd-day-of-occupynigeria-the-police-came-to-attack-innocents/. Last accessed June 13, 2013.

5. An executive member of JAF said the following about the role of social media: "beyond a wider and quick access to information, social media, in the context of the January uprising wasn't the driving force nor constitute the empowerment basis of the critical mass in Nigeria. The critical factor was organisation and the willingness of Nigerians who participated actively on the basis of their resolve to say 'enough was enough'."

REFERENCES

Aihe, Okoh. 2010. "Country Focus: Nigeria is Increasingly Connected." *ITU News*, Jan-Feb. September 12. http://www.itu.int/net/itunews/issues/2010/01/31.aspx.

Aiyede, E. Remi. 2003. "The Dynamics of Civil Society and the Democratization Process in Nigeria." *Canadian Journal of African Studies* 37 (1): 1–27.

Akanle, Olayinka. 2013. "The Development Exceptionality of Nigeria: The Context of Political and Social Currents." *Africa Today* 59 (3): 31–48.

Banda, Fackson, Okoth Fred Mudhai, and Wisdom J. Tettey. 2009. "Introduction: New Media and Democracy in Africa—A Critical Interjection." In *African Media and the Digital Public Sphere*, edited by Okoth Fred Mudhai, Wisdom J. Tettey and Fackson Banda, 1–20. New York: Palgrave MacMillian.

Bastian, Misty L. 1999. "Nationalism in a Virtual Space: Immigrant Nigerians on the Internet." *West Africa Review* 1 (1). http://www.icaap.org/iuicode?101.1.1.2.

Bohman, James. 2010. "Expanding Dialogue: The Internet, the Public Sphere, and Prospects for Transnational Democracy (2004)." In *The Idea of the Public Sphere: A Reader*, ed. Iostein Gripsrud, Hallvard Moe, Anders Molander and Graham Murdock, 247–269. Maryland: Lexington Books.

Cohen, Julie E. 2007. "Cyberspace as/and Space." *Columbia Law Review* 107 (1): 210–256.

Democratic Socialist Movement. 2002. *Nigeria: Civil Rule in Danger*. SocialistAlternative.org. Accessed August 12, 2013. http://www.socialistalternative.org/literature/nigeria/.

DiMaggio, Paul, Eszter Hargittai, W. Russell Neuman, and John P. Robinson. 2001. "Social Implications of the Internet." *Annual Review of Sociology* 27: 307–336.

Everett, Anna. 2009. *Digital Diaspora: A Race for Cyberspace*. New York: Suny Press.

Filipiak, Agnieszka, and Eliza Kania. 2013. "The Precariat: Between Revolution, Temporary Protest or a Message for Change: Conceptualising Precarity in Light of the Occupy and Indignados Movements." *R/evolutions* 1 (1): 18–21.

Ford, Jolyon. 2012. *Democracy and Change: What are the Prospects for an "African Spring?"*. July 14. Accessed September 19, 2013. http://forums.ssrc.org/african-futures/2012/07/14/democracy-change-prospects-african-spring/.

Gboyega, Alex, Tina Søreide, Tuan Minh Le, and G. P. Shukla. 2011. *Political Economy of the Petroleum Sector in Nigeria*. Policy Research (Working Paper 5779), Public Sector Reform and Capacity Building Unit, Washington DC: The World Bank.

Gripsrud, Iostein, Hallvard Moe, Anders Molander, and Graham Murdock. 2010. *The Idea of the Public Sphere: A Reader.* Maryland: Lexington Books.

Index Mundi. 2013. *Nigeria Economy Profile 2013.* February 21. Accessed September 13, 2013. http://www.indexmundi.com/nigeria/economy_profile.html.

Internet World Stats. 2013. *Africa.* September 23. Accessed September 25, 2013. http://www.internetworldstats.com/africa.htm#ng.

Jacobs, Sean. 2013. *New Media in Africa and the Global Public Sphere.* February 21. Accessed August 14, 2013. http://forums.ssrc.org/african-futures/2013/02/21/new-media-in-africa-and-the-global-public-sphere/.

Joint Action Front. 2012. *Joint Action Front (JAF): JAF Deplores the Suspension of the Indefinite Strike/Mass Action.* January 16. Accessed June 2, 2013. www.jointactionfront.blogspot.com/2012/01-jaf-deplores-suspension-of-indefinite.html.

Joint Action Front (Blog), 2011/2012. www.jointactionfront.blogspot.com

Kennedy, Michael D. 2011. *Global Solidarity and the Occupy Movement.* The Social Science Research Council. December 5. Accessed August 13, 2013. http://www.possible-futures.org/2011/12/05/global-solidarity-occupy-movement/feed/.

Lewis, Peter M. 1994. "Endgame in Nigeria? The Politics of a Failed Democratic Transition." *African Affairs* 93 (372): 323–340.

Mayes, Warren Paul. 2009. "Unsettled Post-Revolutionaries in the Online Public Sphere." *Sojourn: Journal of Social Issues in Southeast Asia* 24 (1): 89–121.

Mazrui, Ali, and Nirvana Tanoukhi. 2011. "Arab Spring and the Future of Leadership in North Africa." *Transition* 106: 148–162.

NBS. 2012. *Nigeria Poverty Profile 2010.* Abuja: National Bureau of Statistics. Abuja

Nigeria. 2004. *Meeting Everyone's Needs: National Economic Empowerment and Development Strategy.* PRSP, Abuja: Nigerian National Planning Commission.

Nwokeji, G Ugo. 2007. *The Nigerian National Petroleum Corporation and the Development of the Nigerian Oil and Gas Industry: History, Strategies and Current Directions.* Joint Baker Institute/Japan Petroleum Energy Centre Policy Report, California: Rice University.

Nwosu, N. I. 1996. "The Politics of Oil Subsidy in Africa." *Rivista trimestrale di studi e documentazione dell'Istituto italiano per l'Africae l'Oriente* 51 (1): 80–94.

Obi, Cyril. 2009. "Nigeria's Niger Delta: Understanding the Complex Drivers of Violent Oil-related Conflict." *Africa Development* XXXIV (2): 103–128.

Obi, Cyril. 2010. "Oil Extraction, Dispossession, Resistance, and Conflict in Nigeria's Oil-Rich Niger Delta." *Canadian Journal of Development Studies* 20 (1–2): 219–236.

Omojuwa, Japheth. 2013. Online interview.

Petras, James. 2011. *Egypt: Social Movements, the CIA and Mossad.* February 15. Accessed August 13, 2013. http://canadiandimension.com/articles/3725/.

SPARC. 2010. *Using Mobile Technologies for SPARC Working Paper.* Working Paper (Updated Nov. 2010), Abuja: State Partnership for Accountability, Responsivenes and Capability (SPARC). Accessed September 2013. http://library.sparc-nigeria.com/Delivery.aspx?AttachmentID=bfd120ee-f4ce-11e0-ab5c-00221964e37f.

Taiwo, Rotimi. 2010. "'The Thumb Tribe': Creativity and Social Change through SMS in Nigeria." *California Linguistic Notes* XXXV (1): 1–18. Accessed August 13, 2013. http://hss.fullerton.edu/linguistics/cln/W10PDF/Taiwo-ThumbTribe.pdf.

Wilson, John. 2012. *One to Another: The Occupy Movement Challenges the Media.* January 23. Accessed August 12, 2013. http://www.opendemocracy.net/john-wilson/one-to-another-occupy-movement-challenges-media.

Zuquete, Jose Pedro. 2011. "Another World is Possible? Utopia Revisited." *New Global Studies* 5 (2). http://www.bepress.com/ngs/vol5/iss2/art3.

Chapter 4

The Internet and Democracy in Arab Spring Morocco

Opportunities and Challenges

Abderrahim Chalfaouat

In the persistent struggle over freedoms, equality of opportunity, dignity, and development, different sociopolitical actors have sought access to channels that would amplify rallies and communicate perspectives. As a key channel, especially in twenty-first century Middle East North Africa (MENA), the Internet has been positioned at the heart of grassroots advocacy and political interaction. With its innate cascading interactivity, the Internet has catalyzed political transition and tied change efforts to the urgent demands of eager populations, especially when their zeal for democracy is fueled by technology savviness, young age, and increasing access to education. Thus, online activism that challenges state control has transmogrified not only political squares but also the perception of the Internet itself that shifted from military and educational beginnings to seeking better living conditions.

Despotic governments have often reacted to the democratizing facet of the Internet via surveillance, blockade, and activist imprisonment. Their coercive response is usually aggravated by such structural hindrances as illiteracy and poverty. Yet, virtual crackdown measures vary contextually. As the Arab Spring has exemplified, Khamis et al. (2012) argues, social media as "catalysts for political change and mobilizers for political action must be contextualized within the broader political and social structure in each country, with all their respective complexities and unique qualities." That is, the cultural and technical infrastructure in each country partakes in determining social media effect and democratic outcome.

Peter Dahlgren (2002) stresses that idea when he explains that "the conditions and circumstances of communications play an important role in perpetuating but also problematizing the established political patterns, as well as in facilitating the alternative developments." That is, social media own considerable potential to broaden or eternalize the status quo. Meanwhile, for politicians, academia, and civil advocacy, social media can problematize social fabrics, ignite democratic mobility, or pave the way for socio-political change. The difference lies in the demographic, economic, and social realities of each political ecology as well as the ways different actors manage to exercise power over the Internet communicative abilities.

So, hegemony and advocacy conflicts characterize communication spheres and devalue deterministic understanding of media roles in democratization. The public psyche and socioeconomic context, rather than technology, determine the democratic development. Given the paramount importance of the context, especially in undemocratic or quasi-democratic systems, new media roles "may be hampered by a culture of secrecy or corruption, institutional resistance, or a lack of technical and institutional capacity," as the UNESCO media development indicators explain.

But for Internet users, new tools help overcome propaganda, monitor public policies, and orchestrate activities to block governments or oust regimes. With the arrival of Web 2.0, communication channels proliferated, awareness-raising accelerated, and government criticism became harsher. As expressed via street marches in 2011, online democracy stretched to "go beyond freedom of opinion and expression to include areas such as democratic media governance, media ownership and control, participation in one's own culture, linguistic rights, rights to education, privacy, [and] peaceful assembly," as the Centre for Communication Rights website explains. User-generated online content allowed publics, in Morocco as in Egypt, Tunisia, Yemen, or Kuwait, to vet official stories, since the victim becomes a journalist or an information source for other journalists to spread the story immediately. Consequently, attempts to alter the Arab Spring into a temporary vogue rather than a history turning point shrank, at least at the media level. Both for Internet-users and virtual police, truth production, sharing, and scrutiny acquire new dimensions since bloggers can "tally offenses and delve deeper." Today, citizens "with first-hand knowledge of the story in question might step forward with photos and videos to contradict the invented details," as Craig Silverman (2012) puts it in his blog post.

In the Arab Spring context, the absence of media freedom and democracy motivated the struggle for genuine public sphere democratization. Social media in tandem with broadcast media, especially Aljazeera, initiated bottom-up pressure to topple despotism and totalitarianism in different Middle East North Africa (MENA) countries such as Tunisia and Egypt. The media combination helped both to spread advocacy news and isolate state propaganda.

In the Moroccan experience, street marchers in 2011 did not call for regime ouster. Yet, they demanded political change via orchestrating social media platforms, news websites, and international broadcast channels. Therefore, the interaction between the Internet and democracy in Morocco, especially with the influence of activism in neighbouring countries, deserves more scientific study. The aim is to shed light on the similarities and differences in Arab Spring youth activism, focusing on how the revolutions have empowered Islamists, notwithstanding the outcomes of the situations in Egypt, Libya, Syria, or Yemen. Also, the situation of the Internet in Morocco will be explained, as well as the ways it has encouraged Moroccan youth to combine virtual and street advocacy. This chapter equally analyzes the local and international legal framework that organizes the relationships within the Moroccan virtual sphere. It ends with an exploration of expectations, changes, and challenges in forthcoming efforts at democratizing the public sphere such as the access to information law.

PROXIMITY SYNDROME

The revolutions that erupted in Tunisia and other Arab countries in 2011 have left their own traces on the Moroccan political square. On February 20, a youth movement took to the streets of Rabat, the capital, in line with increasing online criticism against foundering public policies. Orchestration of virtual activism and street marches has been termed the February 20 Movement (M20F) ever since. Under this umbrella term, youth from different economic layers, representing sometimes conflicting sociopolitical projects could gather. A mosaic emerged combining few Party of Justice and Development (PJD) leaders, some Salafists, and "the Islamist-oriented Justice and Charity Association (Aladl Walihsan is also referred to in the literature as Justice and Spirituality Organization (JSO), which this paper adopts), labor unions, and leftist organizations to create a mass movement demanding political change," as Hendry (2013) argues. Their nationwide street marches were initiated, stirred up, orchestrated, and covered online.

Street protests and sit-ins were not the only proximity effect the Arab Spring left on Morocco. In pre-2011 Arab world, severe socio-economic conditions, soaring levels of corruption, illiteracy, and lack of opportunity were not unique to a single country, though at varying degrees. Consequently, Moroccan dissidents called for not only bread and butter, but also equal opportunity, wealth redistribution, dignity, freedom of expression, and more democratization, resembling Tunisian, Egyptian, or Yemeni advocacy in 2011. In addition, various aspects of net activism took place in Morocco across age, gender, education, socioeconomic, geographical, and cultural layers.

Chief among the similarities was the presence of youth. Prior to 2011, the Arab Media Outlook 2008–2012 noted that "over 50% of the population in Yemen, Oman, Saudi Arabia, Jordan, Morocco and Egypt are estimated to be currently less than 25 years old, while in the rest of the [Arab] countries the under-25, 'net generation' makes up around 35% to 47% of total population," as Noman (2015) puts it. For the report, the growing number of youth was indicative of a thriving market for Internet companies as well as a growing need for Internet facilities and projects. Politically, it also meant growing public pressure to demand better socioeconomic infrastructure and more rights.

Another Arab Spring common feature was the absence of traditional leadership. Protest groups were diverse and led by many, but very few activists could claim leadership or exclusive representation, or assume total responsibility for uprising outcomes. Possibly, the Internet contributed to such leadership profusion—or confusion—due to freedom of online assembly and the abundance of grouping possibilities.

Moreover, the use of new media and mobile phones to take instant shots was another Arab Spring affinity. Different youth movements in Morocco, for example, (M20F, Baraka, Mamfakinsh, Moroccan youth against corruption and despotism, etc.) utilized social networks to organize weekly demonstrations nationwide while videos were regularly uploaded to YouTube, some of them taken by phone cameras.

A third aspect was continuous communication with Aljazeera. The role Aljazeera played was two-fold. Moroccan activists, similar to other activists in the region, sent videos to http://sharek.aljazeera.net/, a video- and photo-sharing platform that went live in 2012. They also frequently participated in Aljazeera Maghreb news programs.

The public MENA intimidation that resulted from successive Western interventions in Palestine, Iraq, Lebanon, or Afghanistan also lay behind the Arab revolutions. As stated by Mouncef Elmarzouki, Tunisian ex-president, the Palestinian intifadas ignited the Arab Spring and functioned as its early model, considering the rest of Arab revolts as "learners from the Palestinian people," as Malek Yassine reported in a post to El-balad on September 9, 2012. That is why the Arab blogosphere heated during Israel's *Pillar of Cloud* assault on Gaza in 2012, as Moghul (2012) highlights. In different street marches, including in Morocco, large populations yelled the same slogan: "the people want to free Palestine." The slogan was raised during the May 1, 2011 Casablanca march, deemed the biggest Arab Spring-related gathering.

Yet, Moroccan activism was different in the roof of demands. As an Arab Spring brand, the slogan "the people want to oust the regime" was rarely chanted in Moroccan streets, unlike in Cairo Tahrir Square or Tunis Freedom Square. The alternative in Morocco was "the people want to topple despotism and corruption." Protests occurred in more than 100 cities and towns, with increasing intensity, similar to the mounting strength of the Tunisian revolution that started with asking for democratization but ended with asking the ex-president Ben Ali to depart. The Moroccan regime tried to avoid a similar fate by introducing legislative amendments to the constitution. After the royal speech on March 9, 2011, a committee was formed to receive suggestions from different parties, unions, and nongovernmental organizations (NGOs).

Yet, similar to their peers in neighboring countries, youth movements excelled at protests more than at suggesting alternatives. M20F refused to suggest amendments, basing their refusal on the committee's—and hence the constitution's—lack of legitimacy, since its members were king-appointed, not publicly elected. Meanwhile virtual activism continued, and calls for mass gatherings, especially in crowded areas of Casablanca, further pressurized the regime.

Another key similarity in the Arab Spring is the empowerment of Islamists. In Morocco, not all Islamists participate in party-based politics, nor do they all believe in ballot box democracy, aka boxocracy. The spectrum of Islamists includes Movement for Unification and Reform (MUR) that backs the PJD, in addition to JSO and different Salafists groups. They took different stances concerning the Arab Spring and the role the monarchy should have in post-2011 Morocco, but they all tried to push the limits for their freedom and visibility in the public sphere. For instance, MUR, together with the PJD and other NGOs, refused to join M20F, but simultaneously launched an initiative they labeled "a call to democratic reform," as a "third way" for Morocco to concretize promises of democracy and power-sharing without resorting to regime ouster. JSO, on the contrary, strengthened M20F, but no definite agreement was reached with leftist participants on whether to clearly call for regime ouster. A Spanish-like constitutional monarchy, real redistribution of powers, and an evener sharing of bounties were called for, instead. However, when the JSO realized that their hopes were unattainable, they withdrew from M20F. The lack of prior agreement on the ultimate goal of demonstrations resulted in disagreement within M20F. Meanwhile, the Internet was an ongoing door for MUR, the PJD, and JSO to share viewpoints and address messages to internal and external public opinions, especially through their official websites, official Facebook pages, and op-eds their leaders share on news websites. With the Internet open nature, JSO could transcend the media embargo the regime imposes on their activities. Equally, the virtual sphere has furthered possibilities for MUR and PJD both to keep adherents updated of national board decisions and to attract more sympathizers.

The Arab Spring has equally solidified the position of Salafists in the Moroccan political fabric. In 2011, fourteen Salafist prisoners of the "War on Terrorism" received a royal pardon, including three prominent figures.

Ever since their capture after the May 16, 2003 attacks in Casablanca, the Internet has served as a platform to call for their release, despite the difficulties of the mission. Furthermore, Mustapha Elhassnaoui, a journalist who used to report on defend and Islamist prisoners, was sentenced to three years in prison over allegations of having links with Moroccan Jihadists in Syria, as commented on alquds.co.uk on September 23, 2013. His capture occurred immediately before he embarked on an investigative report on Moroccans who join the Islamic State (IS) in Syria. Yet, some of the released leaders have joined Renaissance and Virtue Party (RVP). The move is another striking outcome of the Arab Spring. It was feared that the Salafists were being co-opted by the regime and exploited by the RVP to infuse new blood in its decaying veins. The move generated discussions between RVP Salafists and other Islamists on Facebook and e-news websites, with open interpretations and online comments from Moroccans at large. The Internet allowed the public opinion not only to follow the debate but also TO contribute to it via opinion articles and discussions on the Facebook walls of different Islamist leaders.

Furthermore, the wind that brought Islamists to power in Egypt, Kuwait, or Tunisia blew westward to enable PJD's landslide win on November 25, 2011. The party gained 107 out of 395 parliament seats. A new government was elected, with Abdelilah Benkiran, the PJD secretary general, as its head.

Shortly after nomination, a number of alterations have occurred to concretize the constitutional prerogatives of participatory public policies and the promises of the Arab Spring. When decisions that necessitated public explanation were taken, Benkiran would address the Moroccans via public TV channels, but the videos would immediately occur on e-news websites and social media. They often spark public debate, indicating a general change in citizen-politician relationship. Youngsters who used to occupy themselves with sensuality and trivia are now more concerned with the future of the country and usually try to evaluate public policies and upload materials for or against the regime. Feeling their mounting importance, the ministry of communication invited around 500 representatives of more than 250 e-news websites to a public hearing. Also, a special session was held to listen to moderators of e-news sites while preparing a white book to organise the field of e-journalism.

In short, different Islamists have benefitted from the winds of the Arab Spring and they have all utilized the Internet to communicate their social projects. Besides, they all avoided calling for regime ouster, but they kept faithful to their agendas. Some opted for "the third way" of change within stability while others chose calling for a constitutional monarchy, within stability too. The two paths presumably emit similar results of seeking reform without sacrificing stability, especially in light of the quagmires that Libya, Syria, Yemen, and Egypt have sunk into.

WEB 2.0: REASONED ATTRACTIONS

Different new media have functioned as catalysts for political change in the Moroccan public square. Principally, Facebook and YouTube in addition to news websites such as Hespress.com, Lakome.com, and Alraiy.com have provided opportunities for freedom of expression and allowed youth to both keep abreast of political changes and participate in political advocacy. In addition, the Moroccan blogosphere (aka blogoma) rose considerably to express youth dissatisfaction with the long-standing corruption and despotism in Moroccan politics, while youth podcasts further blurred the lines between the virtual world and seeking freedom of speech, effective socio-economic policies, and evener wealth distribution.

The role and infrastructure of social media have changed rapidly. A Ministry of Communication report on media changes in 2012 considers the expansion of Net connectivity and penetrability an indicator of freedom of the press. It says that about 51% of the population (or 16.5 million) have access to the Net. It also posits that Morocco leads the Arab world and Africa in bandwidth, infrastructure, and legal framework. "ICT growth was largely dominated by a 553 per cent increase in 3G internet services provided between June 2007 and June 2008. In fact, 3G internet services accounted for 24 per cent of the total market share, compared with 74 per cent for ADSL broadband Internet," as the OpenNet Initiative listed on its website on August 6, 2009. Moreover, International Telecommunications Union (ITU) statistics inform that Morocco was home to 16,477,712 Internet users on June 30, 2012, making up 51% of the whole population, compared to 15,656,192 in 2011, 4,600,000 in 2007 and 10,442,500 in 2009. Apparently, numbers peaked in 2011, explained by the growing interest in different Internet facilities, the widespread installation of Internet cafés, and the personalized connectivity that 3G wireless connection offers. Compared to other African countries, Morocco ranked third in the list of the top ten, internet users, according to Internet World Statistics for the second quarter of 2012.

Internet statistics for Facebook also ran high for Morocco in 2011, to rank third in the MENA region. Between February and June 2011, the number of Moroccan Facebook users "rose by over 600,000 users to a total of 3,525,460 users. In June 2011, Facebook was the most frequently searched and visited site in Morocco, ranking higher than the Google search engine," as Rahman (2012) explains. News sites have also proliferated from around 500 to 700—at least the ones the Ministry of Communication could count—in 2013, to make of the Internet a serious watchdog for human rights abuses, shaky public policies, or traditional media blackouts. They present opportunities for freedom seekers who dream of a country on the move toward reduced degrees of corruption and despotism.

The Internet has attracted Moroccan surfers for a plethora of reasons. First, it appeals to the interests of both the educated and the illiterate, especially given the stark levels of illiteracy in Morocco. Morocco's newspaper readership is one of the lowest in the Arab world. According to the Arab Press Network figures, for Morocco's "33,8 million inhabitants, all print media combined sell 350,000 copies/day, compared with 1,3 million copies in neighboring Algeria, which has 33,3 million inhabitants," as a Maghrebinenglish blog listed. Put differently, sales stand at thirteen copies for "every 1,000 people (the worldwide average is 250 per 1,000 people; in Arab countries, the average is 40 per 1,000 people and in Algeria, the average is 71 per 1,000 people). Only 350,000 copies of newspapers are distributed every day," as Lmaati Monjib (2011) puts it. The government tries to intervene, but in vain. The first report on freedom of the press that the Ministry of Communication issued in 2013 informs that "65M MAD [$7.6M] was dedicated to subsidizing the print press in 2012." Notwithstanding, newspaper readership shrinks increasingly.

Second, the Net's participative and interactive features help transcend broadcast media's corruption and denigrating content. Fatima Elifriqui, TV art and cultural programs anchor, said that the main challenge that the new government's media reforms may face is media lobbies. In an interview with Wijhatnadar.com (points of view in Arabic), she argues that the current communication ministry is helpless toward TV lords since they are appointed by the king who holds key power over the media, including the authority to appoint the heads of all public radio and television stations. So, according to her, little can be done to change traditional media chaos, since "one can become a billionaire in a month." Possibly, this is another characteristic the Moroccan mediascape shares with the rest of the Arab world, though some committees have been set to curb corruption and democratize the selection of TV programs for public funding.

In a nutshell, the expansion of Moroccan navigators, youngsters more particularly, suggests that they have come to the understanding that the media, as an influential determinant of relations in the public sphere, vividly contributes to the demarcation of powers, challenging consent and triggering protest. Somehow, democratizing access through the low prices initiates the Moroccan society into a knowledge-based fabric. Today, youth audiences are more aware of the ways undemocratic regimes have been managing minds. Unsurprisingly, TV and radio continue to spread official stories and one-dimensional interpretations of political reality. Thus, the strategic role of traditional mass communication is neither deniable nor contestable, but has become challengeable by new modes of information and standpoint sharing.

THE NET AS CATALYST OF ACTIVISM

The Internet offers Moroccan youth an alternative public sphere for freedom of expression. For instance, street activism was streamed live for Moroccans inside and outside of the country. Surfers also managed to organize group discussions, make sarcastic Photoshop caricatures, write mocking Facebook statuses and share podcasts. For instance, an unposed shot was taken of member of Parliament (MP) displaying her legs loosely, maybe due to tiredness. Soon, much debate rose about her quasi-naked thighs. Though the issue sparked and circulated on the Internet, MPs applauded to their colleague in an open parliamentary session encourage her against the sensitivity of the photo, and openly sympathized with her right to privacy, indicating that they followed the developments online.

Internet activism has influenced democratic practices in Arab Spring Morocco in three main ways: (a) drawing attention to uprisings; (b) monitoring public policies; and (c) pushing the limits of freedom of expression.

A sound example in the first category is the case of Taza demonstrations. On February 1, 2012, the police and protesters in Elkoucha neighbourhood clashed. The Internet was central in unfolding the different facets of the story. While protesters, reportedly OSJ adherents, uploaded videos of police brutality, other videos were shared of stone-throwing protesters who blockaded the streets. Local news sites kept updating the public with development stories. Also, a Facebook page was launched to share more materials. The high circulation pushed local MPs to visit the city and try to settle local problems of unemployment, infrastructure, and basic services, though activism seemed to have political underpinnings too. An official committee was founded to concretize the promises while a human rights NGO started its own investigation of the alleged violations.

Another outbreak caught Marrakech in fire in January 2013, despite its touristic reputation. Protests objected to soaring electricity bills, and the police reacted violently, using tear gas, water trucks, and batons. The Net allowed protesters to foreground the uprising and transcend traditional media blackout. Videos reached wide viewership via YouTube and local, national, and international e-news websites.

Furthermore, on February 16, 2013, the vice-solicitor general in the city of Midelt obliged a mechanic to kiss his shoes in a police station. His only "sin" was to ask the judge to queue to get his car repaired. The vice-solicitor refused to wait and sent two policemen to summon the mechanic. After the release of the intimidated young mechanic without charges, the small city of Midelt was in an uproar against the humiliating treatment. Immediately, the news spread nationally via social media and e-news websites, especially when a sit-in of more than 5,000 angry demonstrators was organized on February 20, 2013. A committee from the Ministry of Justice and Public Freedoms was mobilized to question the judge. Labelled "the shoe owner," he was stopped from work for six months as a punishment and has been transferred to another city, to mark another case whereby social media amplify street activism and help spotlight despotic offences.

Secondly, the Internet hosted different discussions of public policies when suitable events erupted. For instance, when a Moroccan youngster claimed to create a plane out of scratch, new media rushed to push the government to encourage him. Nicknamed "the plane owner," his claims lacked a solid ground, but the Internet mobilized the Ministry of Transport to send a committee to meet the ambitious youth and discuss ways to improve his plans. It was also an opportunity to discuss public policies of supporting creativity, instead of squandering public funds on nationwide music festivals. Moreover, the public policies of scientific research and the meager finances allocated to the domain were discussed.

A second example is the case of Amina Elfilali, a 16-year-old girl who "committed suicide" after being obliged to marry her rapist. Immediately, human rights activists, especially feminist ones, brought into public attention the policies of forcing minors into marriage. Different sit-ins were organized in front of the Parliament—with videos and photos widely shared—in tandem with lengthy discussions online and on traditional media to call for legislative changes, especially item 475 of the penal code that permits rape impunity through marriage, notwithstanding the age of the raped, as Elachraf Hassan reported on Alarabya website on March 8, 2012.

A third example occurred, when, in the middle of the heat of 2011 demonstrations, a terrorist attack was waged on the Argana café in Marrakech. The instance presented an opportunity to discuss the timings of terrorist blasts in Morocco, especially since Al-Qaida in the Islamic Maghreb (AQIM) did not claim responsibility. On the contrary, AQIM publicized a note denying responsibility and asking Moroccans to look for the perpetrator elsewhere. It was also an opportunity to question the public policies of terrorism-fighting and the success of the security approach.

The third categorization, extending the limits of freedom of expression, manifests in such activism as Mounir Agueznay's sniping or Nafissa Elhouari's podcasts. Mounir, aka Targuist Sniper, from the town of Targuist in the northern district of El-Hoceima, started a long journey of exposing police corruption. In 2007, he uploaded four videos of car drivers bribing policemen, which resulted in both copping with gendarmerie and encouraging Moroccan youth to disclose more corruption and bribery cases.

In addition, Nafissa podcasted just three episodes of her YouTube program "Bitter Honey" before local authorities started to investigate her. The label "Bitter Honey" is indicative of the contradictions of Moroccan politics whereby "honey" stands for Morocco, a valuable substance, while "bitterness" stands for the "sourness" resulting from social, political, and economic hardships Moroccans undergo. Her three videos dealt with "the Internet" and how slow it is, "Crime" and how police absence aggravates its outcomes, and "the Mawazine Festival" and how it squanders public funds. She relinquished podcasting when the police harassment started. The third video, in particular, was largely visited on YouTube (257,987 views) and shared widely on news media. Her bold criticism of "the Mawazine Festival" contributed to strengthening street protests against it.

THE LEGAL FRAMEWORK

The economic, cultural, and political potential of the Internet in Morocco puts challenges before legalizing the incessantly renewed virtual tools and encounters. That is why the Moroccan government initiated efforts to meet the benchmarks of the Millennium Global Goals. Starting in Tunis 2005, the goals encouraged governments to connect villages, health centers, educational institutions, libraries, government agencies, and facilities to the Net. Even though the neoliberal aspect of such connection can facilitate unplanned intrusion in a globalized world, Morocco is not lagging behind on WSIS broadband, accessibility, and e-government benchmarks.

The stocktaking report of 2013 cites a number of e-projects in Morocco. Two platforms (maroc.ma and service-public.ma) have been established to meet the promises of e-Morocco—a national strategy partnering private and public stakeholders. Also, the National Telecommunications Regulatory Agency (ANRT) launched a Programme for Generalized Access to Telecommunications (PACTE) "to ensure access to telecommunication services for the entire Moroccan population and, more specifically, to provide internet and telephony to 9275 localities identified as dead zones." The programme, which incorporates the technologies of GSM, EDGE, VSAT, and 3G, was adopted in November 2006 by the Universal Service Management Committee (CGSUT). A total of 8,046 localities were covered by 2012. Besides, the Ministry of Interior has set up an electronic desk—Watiqa.ma, which allows users to apply for administrative documents virtually and receive them by mail. The service currently is available to order birth certificates or a full extract from the civil registry. In January 2012, a first attempt started in Rabat, the capital. Forthcoming steps are to cover the whole country and include other administrative documents. In addition, administrative procedures are facilitated via the IDARATI program.

While digitization projects indicate a willingness to democratize information, resolutions are equally taken to curb freedom of expression. Law 33–03, or "anti-terrorism code," is the first legal text to clearly sanction cybercrimes, and prohibit any use of Internet data and information circulation that can be interpreted as a terrorist crime. In 2009, Morocco passed on an Internet law to protect personal data and ban unauthorized access to private information and its treatment in any form. Meanwhile, Morocco is a signatory of a number of international Internet policy agreements. It was among the eight early countries to sign on ACTA, given the controversy of the act and its repercussions on freedom of expression and communication worldwide. These laws combined blur the image of what a cybercrime is since they allow Internet service providers (ISPs) to surveil activism and permit Internet police to deal with personal data without much precaution, especially under pretexts of fighting terrorism.

OFFICIAL REACTIONS TO ACTIVISM

True, YouTube and other social media platforms contribute vividly to publicizing, promoting, or organizing democratization activities. Nevertheless, indications of Internet surveillance creep up every now and then.

Given the economic benefit the Internet generates for service companies and the quest to bring more freedom of expression into the political sphere, surveillance measures in Morocco are neither absent not systematic. That is why, activists, journalists, bloggers, and social media users rarely question the technologies specialized security services use to catch them red-handed criticizing the regime. Yet, whenever the limits are transcended, the stick approach reigns supreme. The limits, or taboos, in Moroccan politics are understood to be: (a) criticisms of the monarchy, especially on human rights abuses; (b) territorial integrity—especially Moroccan Sahara policy issues; (c) religion; and (d) corruption in the military.

Street interventions against activists coalesce with ones waged on the Internet. In 2007, YouTube was banned for a few days, but state-controlled service provider Maroc Telecom blamed the problem on a technical glitch, not in reaction to videos that document abuses in the Moroccan Sahara. A first victim of Internet surveillance was an IT engineer, who received a three-year sentence and a $1,350 fine in 2008 for creating a fake Facebook account with the name of Moulay Rachid, the monarch's brother. The international online and offline sympathy Fouad Mortada received called the regime's attention to the new challenges of the Internet and how difficult it had become to keep human rights violations unnoticed. Fouad later received a royal amnesty. A few months later, Mohammed Erraji, was arrested for writing a blog post that was deemed disrespectful to the king, but he received a royal pardon too, when national and international condemnation increased, as Jankari Rachid commented on Menassat on September 12, 2008. On the contrary, Haydour, a student who denounced the monarch in a YouTube video, received no royal amnesty for his three-year sentence because the insults he levels at the monarch seem harsher than tolerable, as a Human Rights Watch comment posited on July 6, 2013. Another example is a caricaturist who, in April 2011, shared photos of the monarch's sister in her bikini while enjoying her vacation on Mexican beaches. Two days later he was arrested, but due to the politically tense atmosphere of M20F protests, he was released. Moreover, Mouad Belgouat, aka Elhaqed, a rap singer, was arrested on March 29, 2012, and received a one-year sentence for distributing invitations to an M20F gathering, but his rap songs bluntly stigmatized socio-economic inequalities and corruption that overwhelms state departments. His lawsuit file included denigrating state symbols when he produced and uploaded a video on YouTube entitled "State Dogs," referring to the monarch's circle and police departments.

Also, when shrill criticism is systematically addressed to the deep state, the terrorism law may be resorted to. Ali Anouzla, editor of lakome.com, wrote ground-breaking articles accusing deep state figures of continual despotism, demonstrating that the regime simply stooped to the Arab Spring storm to come back after public nerves calmed down. He was first to bring attention to the royal pardon that Daniel Galvan, a Spanish paedophile, had received. The scandal triggered unprecedented activism against royal decisions. Despite his leftist inclinations, Ali was sued under the terrorism code for encouraging terrorism, when his news site lakome.com quoted a hyperlink to an AQIM video. The national and international condemnation of his arrest led to his release after spending a few weeks in custody.

The problem with these cases is that imprisonments resort to the penal code. They equate intellectual opinion expression to burglary or street aggression. Very often, security services use more logical pretexts to legitimize the arrests such as threatening political stability, promoting terrorism, spreading blasphemy, or even committing adultery. Consequently, when the trumped-up charges are too naive to be acceptable or credible, the indictees are granted pardon. For the public, the open, decentralized, and interconnected nature of social media has made such prison-leading surveillance obsolete and outdated. Therefore, changes in legislature and limiting the omnipresence culture of the Ministry of Interior necessitate more policy advocacy for an opener Internet. Otherwise, Morocco will continue to lag behind in Internet freedom reports and in providing Moroccans with the Internet they deserve.

CIVIL SOCIETY ADVOCACY

It seems an epistemological characteristic of revolutions that political and cultural differences shrink while activism heats up. This manifested in Internet behaviour of Moroccans while street protests peaked. Most Facebook pages and news websites' opinion articles focused on common concerns, especially ousting despotism and corruption. This led to a high level of tolerance and coexistence. When calmness reigned, diverting trails regained power. Even though dissident cohorts still seek more democratization, when they took diverting trails within civil society NGOs, their activism abated.

Therefore, general dissent occurred when petulant youth took to the streets to denounce two royal moves. The first was to protest against the humiliating annual "loyalty feast," whereby MPs, ministers, NGO representatives, and city notables bow down before the monarch to renew loyalty. The controversial bowing has often been denounced, without managing to cancel it. Protesters start calls online and share videos before they sit-in against the degrading protocols, but the police often intervene, sometimes violently such as on July 30, 2012.

The second was protesting against granting a royal pardon to Galvan, the Spanish pedophile. Galvan had raped 11 Moroccan kids in Marrakech, aged between 14 and 3. He was sentenced to thirty years imprisonment, but was released after barely serving for two. The scandal was primarily uncovered by a news website. Then, sequel to a Facebook call to protest on August 3, 2013, a number of youth, artists, and politicians marched in front of the Moroccan parliament. When beaten, a national agitation erupted via street demonstrations, heated online debates, and foreign TV coverage. Protesters bravely faced police brutality. Consequently, the royal palace withdrew the pardon, a senior prison official was pushed to resign, and the king received the families of the raped kids.

Furthermore, in post-2011 Morocco, cultural debates, both online and offline, seem less tense than before, possibly due to the fact that the new constitution recognizes the cultural diversity of the country and because more light is shed today on counter-revolution tactics to deprive the constitution of its bridging nature towards a more democratic public sphere. Yet, the Internet is often utilized for cultural debates. For example, "let's come to read" was a youth cultural initiative in reaction to the sensuality and money-squandering of the Mawazine Festival. Also, in "the Facebook kiss," a teenager photographed two schoolmates "kissing" and shared the photo on Facebook, unaware of the repercussions in a conservative society, The school was furious when the photo was circulated in the social media. In reaction, a local human rights NGO filed a lawsuit against the three teenagers over allegations of encouraging sensuality. Soon after, the three were summoned by the police, and the "kissing" boy spent few days in custody.

FUTURE PROSPECTS

The discussions in this chapter portray both the potential the Internet holds for Moroccan democracy and the obstacles that hinder an ampler contribution to youth's hunger for freedom of expression platforms. The 2011 constitution recognizes the importance of citizen participation in suggesting, promoting, and evaluating public policies, especially through the mechanisms of civil society. The participative approach is a resultant of demographic changes and the inability of traditional structures to attract youth enthusiasm for political institutions. The government tried in vain to integrate youth through culture. The Internet, on the contrary, offers a considerable margin for freedom and advocacy, but a broader integration necessitates more amendments and openness to insure youth sustainable focus on immediate environment concerns.

To forge a context that parallels the changing and challenging media atmosphere, structures and practices need to meet the ongoing democratization process. Given how civil society has become a fourth estate in the 2011 constitution, different participative democracy opportunities need to consider defending Internet freedom and network neutrality. Incorporating youth civil society in Internet governance can be indicative of a serious tendency to sustain youth participation in the development of the country. That is, multistakeholderism can be viewed as a key step for ensuring communication rights for all and provide a freer Internet that Moroccans deserve, especially when Net-savvy youth are integrated.

Besides, novel changes in media laws in Morocco should address challenges of the intersections of the Internet and freedom of expression. The forthcoming access to information law needs to insure administrative accountability and access to citizen-relevant information, not only by journalists but by the whole population. In addition, the forthcoming press law needs to be distanced from the penal and terrorism codes. Otherwise, the story of allowing the media at large and citizen particularly to suggest, promote, evaluate, and scrutinize public policies won't be granted the laurels of success. In that case, the price is to be paid by both policy-makers and a whole generation that is hopeful of truly developing Morocco. Finally, TV Channel Four, aka the cultural channel, needs to be more effective in reducing the high illiteracy rates, push towards knowledge-based society, and contribute to bridging the digital divide, in line with the nascent media reforms. Its entire program grid should be changed to include the promises digitization holds for the Moroccan public sphere.

To conclude, media are key in hammering a democratic interpretation of the new constitution and sharing change progression with the public. Without their democratization, a considerable portion of the changes the Arab Spring has come to concretize will go astray, and much government effort will be undervalued, which may trigger stronger protests at any time.

REFERENCES

Arab Media Outlook 2009–2013. *Inspiring Local Content, Forecasts and Analysis of Traditional and Digital Media in the Arab World. 2010*. Dubai, UAE: Dubai Press Club.

Axford, Barrie, and R. Huggins. 2001. *New Media and Politics*. London: SAGE.
Bennett, W. Lance, and Robert M. Entman. 2001. *Mediated Politics: Communication in the Future of Democracy*. Cambridge: Cambridge University Press.
Center for Communication Rights. 2015. "Introduction." http://centreforcommunicationrights.org/starting-points/introduction.
Dahlgren, Peter. 2002. "In Search of the Talkative Public: Media deliberative democracy and civic culture." *Javnost*. 9 (3), 5–25.
Hackett, Robert A., and William K. Carroll. 2006. *Remaking Media: The Struggle to Democratize Public Communication*. New York: Routledge.
Hendri, Krista. 2013. "Morocco After the Arab Uprisings: Key Policy Recommendations." The Fund for Peace Roundtable discussion report, Washington. http://library.fundforpeace.org/library/shrtr1320-morocco-10a.pdf.
Howard, Philip N. 2010. *The Digital Origins of Dictatorship and Democracy: Information Technology and Political Islam*. Oxford: Oxford University Press.
Human Rights Watch. 2013. "Despite Reforms, Moroccan Student Still in Jail for 'Insulting the King.'" http://www.ifex.org/morocco/2013/07/16/student_imprisoned/.
Khamis, Sahar, Paul B. Gold, and Katherine Vaughn. 2012. "Beyond Egypt's 'Facebook Revolution' and Syria's 'YouTube Uprising': Comparing Political Contexts, Actors and Communication Strategies." *Arab Media & Society*. http://arabmediasociety.com/articles/downloads/20120407120519_Khamis_Gold_Vaughn.pdf-Monjib, Maâti. 2011. *The "Democratization" Process in Morocco: Progress, Obstacles, and the Impact of the Islamist-Secularist Divide*. Washington, DC: Saban Center for Middle East Policy, Brookings Institution.
Moghul, Haroon. 2012. "From Confusion to Kardashian: Misreading the Middle East." *Religion Dispatches*, December 5. http://religiondispatches.org/from-confusion-to-kardashian-misreading-the-middle-east/.
Noman, Helmi. 2011. "OpenNet Initiative: Middle East and North Africa." http://library.fundforpeace.org/library/shrtr1320-morocco-10a.pdf.
Rahman, Zahir. 2012. "Online Youth Political Activism in Morocco: Facebook and the Birth of the February 20th Movement." *Journal of New Media Studies in MENA*. 1, Winter. https://www.academia.edu/1417051/Online_political_activism_in_Morocco_Facebook_and_the_birth_of_the_February_20th_movement. ISSN 2162–6669.
Yassine, Malek. 2012. "Palestinian President: Palestine is the First Spark of Arab Revolutions." *El-balad*, September 6. http://www.el-balad.com/185557

Chapter 5

Occupy Movements, Alternative New Media, and Utopia

The Gezi Resistance and Activist Citizen Journalism

Barış Çoban

With the coming of the Gezi Resistance, all that was solid in political life melted into air. The Gezi Resistance created a topsy-turvy political milieu and spurred the first steps of a new long march. However, this event is not to be taken not as a sudden, spontaneous uprising, as mere resistance, but rather as a process in which the seeds of a revolutionary transformation were sown. The Gezi Resistance was a "leap," the leap of a tiger that feeds the poetry of the future; the start of a long march.

With the Gezi Resistance, the social opposition in Turkey commenced a process of major transformation, and it may be claimed that this transformation was in the same revolutionary spirit as all the "Occupy" movements, beginning with the Arab Spring, the Spanish Indignados, and the US Occupy Wall Street movement. The Gezi Resistance gave rise to a movement that transcended the usual repertoire of political movements in Turkey and, for this reason, the traditional revolutionary circles initially failed to grasp its nature or act as its vanguard. This new social movement, which completely broke all the old moulds and habits, has continued to prove that an apparently apolitical and passive youth is capable of politicization and mobilization. Furthermore, the establishment of the "Gezi Commune" showed that it is possible to create and sustain a communal living-space in what might be aptly regarded as the heart of the country. The commune experience is a practical manifestation, in the Badiouian sense of the "communist hypothesis/idea," not of a mere search for alternatives but of the creation of these alternatives, and a political rupture breaking with the existing political paradigm.

In this work, the new political formations created with the Gezi Resistance will be examined on the basis of Hakim Bey's concept of the "temporary autonomous zone" and Lenin's "dual power." In the "temporary autonomous zones" (Hakim Bey 1991) such as those represented by the Gezi Resistance and other Occupy movements, which we might term "living utopias" on the basis of their communal structure, a state of "dual power" (Lenin 1974) is created. The creation of a temporary state of dual power forms a revolutionary experiment, and these experiments create cracks in the base of the system which, when these cracks unite, gives rise to a network of revolutionary resistance. In this sense, it is possible to claim that, reading beyond the classical Leninist usage of the concept of "dual power," namely, referring to the stage of revolution prior to the final period, dual power situations may be created in the context of all power struggles. One of the *sine qua non* features in the practice of any temporary dual power situation created on the basis of autonomous zones is the emergence of new alternative media. The way, in which activist citizen journalism[1] was used in the establishment of alternative media in the Occupy movements, particularly the Occupy Gezi movement, as a revolutionary instrument, will be explored according to the conceptual framework of W. Benjamin.

COMMUNIST IDEA AND UTOPIA

The revolutionary process[2] is formed from the synthesis of splits, defeats, and resurgences that take place in the social environment. As well as class conflict, the tension and conflicts occurring between social groups associated with every other social problem and the ruling power give rise to, and are accompanied by, the emergence and rise of social movements.

In connection with their aim of exerting a lasting influence, social movements struggle for the creation of "temporary autonomous zones."[3] In connection with their long-term struggle with the reigning authority, the creation of autonomous zones may be interpreted on the basis of postponing the revolution for the future. Stressing the importance of immediate needs and the assertion that it is possible to commence the process of revolutionary transformation by taking the first steps straight away has begun to make the struggle to create autonomous zones within the social living space even more important. Struggles for autonomy are essentially rooted in space and imply a form of urban struggle. The rural-based counterpart to this, with its struggles to create communal living space, are also important, indeed it is apparent that they represent a success even beyond that of urban experiments, however the power to effect global transformation remains the preserve of urban-centered movements. The establishment of temporary autonomous zones also simultaneously creates a situation of "dual power,"[4] as it is known in Leninist terms. The class holding power in a temporary autonomous zone creates a *de facto* state of temporary dual power, and, in turn, this region becomes a locus of intensified class struggle for the duration of its existence. Although it may appear that the struggles of the occupiers in their battles with the ruling authority to defend a small area, in the context of temporary autonomous zones, are not particularly ambitious, this situation of dual power may, in fact, contribute to the spread of power struggles in localities throughout the country, as well as to both the increase in number and permanence of autonomous zones.[5]

The city is a war zone. Social movements attempting to liberate the city from the grip of the ruling power are carrying out a new kind of guerrilla warfare. This struggle is waged on the basis of avoiding violence as much as possible, with the exception of legitimate self-defense, through active and passive forms of resistance. These new forms of urban-based struggle both contain and transcend the experiments of the past. It is possible to speak of this as a new kind of mobilization, able to operate as important power in the poor areas of the city but also of engagement in the city center. This process enables the urban poor to come together with the new working class that is, proletarianizing white-collar employees, on the basis of networks of resistance. The painful process through which the city's poor and the proletarianized new class come under each other's mutual influence forms the crux of the social movements of the current era. Traditional movements, with their traditional networks of resistance, carry out their struggles by combining their efforts with the digital networks of the new class movements to form hybrid networks.

Struggles based on place show the development of a structure that both contains and transcends the experiments of the past. Beyond the establishment of the neighborhoods of the oppressed as "liberated zones," through the temporary occupation of the city center, or by holding actions and erecting paralyzing barricades, social struggles are spread throughout the city, and certain zones are also opened up. During this process, in which an effective, widespread network, creating a number of revolutionary bases, is developed inside the city, the communication and mutual influence that exists between this network and the new, alternative media networks creates the potential for an effective revolutionary struggle against the ruling power. Until now, one of the basic reasons for dissidents to take to the streets has been to become audible and visible in the face of willful blindness and feigned deafness. In directly opposing the representation-based, anti-participatory political system, the monophony of the mainstream media and, without mediation or representation, the ruling power, the dissidents, struggling with both time and place, create their own participatory-democratic political practices. Moreover, this process also entails the creation of new communicative practices, the formation of new tactics, and the founding of new alternative media channels.

The creation of new media and media channels is born of necessity. In order for dissidents to be able to share news and information during this period, as well as make and implement political decisions and keep the general public informed of their actions, they employ new alternative media and create their own alternative networks. In relation to the network society described by Castells,[6] the process of creating alternative networks runs in parallel with the process of establishing a temporary autonomous zone or a state of dual power. The establishment of an alternative media and alternative networks is a *sine qua non* of successfully creating an autonomous zone or dual power.[7] The Arab Spring and subsequent Occupy movements founded their creation of temporary autonomous zones and dual power on the basis of social media. During this process, the activists both created their own media and made themselves into a form of media.

The struggle for autonomy is actually one of the way stations on the road to revolution and it harbors signs confirming that transformation is possible and that a future overthrow of the entire existing system and its revolutionary reconstruction in the future may arise from what is built on the basis of today's struggles. In this connection, they aim to understand the conditions that make the current practices of present-oriented revolutionary movements possible on the basis of their past experiences and seek a revolutionary struggle able to open a path into the future: "For the work of reconstructing the present, finding its preconditions in the past, projecting its likely future, and seeking out the preconditions of this future in the present, now conceived of as an extension of the past, is never truly finished" (Ollman 2008, 21). This revolutionary process based firmly in the present, first takes a tiger's leap into the past, as the practical experience gained from this leap actually sets the scene for another tiger's leap into the future and renders this possible because the aim, in making a great leap towards the future, is to actualize the experience and foresight of the past. Both of these great leaps are centered on the present, and have been made possible by the autonomous zones created in the present. The struggle for autonomy[8] may be read as a revolutionary form of politics which begins in the present and employs all the means of struggle possible in all living spaces to break up the state into smaller and smaller pieces, create cracks, and subsequently, brings these cracks together to render the system inoperative.[9]

The Gezi Resistance is the linchpin of a political transformation in Turkey, and it has irreversibly transformed the political life of the country. In the wake of the Gezi Resistance, Turkey's revolutionary political movements were openly confronted with their own need to transform themselves. In the context of the permanent or temporary autonomous zones and the formation of a new political culture on the basis of areas of "dual power" created by class-based movements and other identity- or issue-based movements and making them work, the carrying out of a realistic revolutionary struggle is absolutely critical.

The Gezi Commune exemplifies the experience of dual power, and it is possible to read the Gezi Commune as a temporary autonomous zone. Gezi, like the other Occupy actions, was a reflection of the network of cracks created in the authority of the ruling power and possessed a rhimozomatic structure emerging as a network formed at the meeting point of various cracks created by struggles around a number of disparate issues. In this sense, the Gezi Resistance was a process that united defensive and transformative actions in a manner unprecedented in the country's history. Thus, the Gezi Resistance possessed a rhimozomatic structure that brought together disparate struggles such as the demonstrations against the refusal to allow May 1 gatherings in Taksim Square in 2013, the struggles for urban rights against gentrification programs (the struggle for the Emek cinema), ecological struggles (opposing thermic and nuclear power plants and hydro-electric schemes), the lesbian, gay, bisexual, transgender, and intersex (LGBT) struggle, class-based struggles (centered on transformative actions), and struggles for rights and freedoms (centered on defensive actions). With the intense participation of civic white-collar employees and unemployed, the movement was an revolutionary surge risen spontaneously which is against the reigning power, and it managed to create great surges that possessed the whole country in a short time. Surges of movement obtained the involvement of many people in a short time, and traditional revolutionary movements also associated with the movements based on rights and freedom in the beginning. This process actually made the "dual power" probable to experience. As the vanishing point of many fractures Gezi Park was an area where a memorable revolutionary practice was experienced for a fortnight. Although it held the traditional forms of politics in Turkey, the Gezi Commune was a practice that had never been experienced before. The state of "dual power" in the Gezi Commune was cherished of this practice. Becoming the center of the country for a fortnight, the Gezi Commune presented a revolutionary practice which put a pressure on the power. Even though it was the leadership and the spokesmanship of "Taksim Solidarity" which was discussed, it occurred that a life, politics, and campaign going on the basis of collectivity was possible. It was evident that so-called finished barricade struggles had not yet finished. Barricades help draw the borders of the revolutionary center or more properly, "temporary autonomous zone" in sense of symbolically referring to the borders which the government could not exceed. Capitalism maintained by barricades in the midpoint of Istanbul for a fortnight, had no authority. The sustenance of such experience made it obvious that the revolution was not a dream to be put off and that the communal life was not a utopia beyond reach. A Communist idea—in Bauman's (1976) term "the active utopia"—became "the event" in the Badiouian sense, and it had existence.[10]

One of the signs of dual power in the Gezi Resistance was the barricade. The barricades constituted the borders of temporary autonomous zone built up in the city center: "In the history of social movements, the barricades erected in the streets had very little defensive value . . . but they always defined an 'in and out', an 'us versus them', so that by joining an occupied site, and defying the bureaucratic norms of the use of space, other citizens could be part of the movement without adhering to any ideology or organisation, just by being there for their own reasons" (Castells 2012, 10). For standing behind the barricade actually meant standing against the power by demolishing all other factions, it made the collective of crowds possible regardless of a number of differences. Barricade struggles gained validity, although it was in a symbolic sense. The revolution of 1848, the Paris Commune of 1871, and the rebellions of 1968—their traditions and their traditional forms of struggle resurrected. "The barricades have come back," as Scholl said (2012). Besides, the barricade struggles, which had been already carried on by the revolutionary organizations in poor districts, moved to the city center. Barricades also are a sign of a sentimental break beyond forming a boundary between the powers. Barricades can be described as the crucial points that connect the past, present, and future as physical symbolic signs of interactivity in historical context. They have the ability to transfer morals created now to the future as they carry all the inheritance of the past to the present. In context of the Gezi Resistance, barricade struggle actually refers to an area where the utopias were realized, all the power of the government was lapsed, and the reigning power was neutralized, which was why it was proved that it was possible to build a communalist life in the area preserved and drawn by barricades in this course, although it was for a short time.

Occupy movements caused a communalist break by cutting the occupied areas off from usual time and place and by creating a new network of social relationships. As to Benjamin, "while play involves a form of mimetic faculty, which liberates the restricted and alineated human sensorium, space entails a material condition for corpereal practice. In the Commune, Benjamin finds a alternative mode of the communal social realiton against the phantasmagoria of the isolated private citizen" (Kang 2014, 193). In other words, it refers to the practice of creating alternative counter public sphere/s (Fraser 1990; Negt and Kluge 1993; Hansen 1993) against the bourgeois public sphere (Habermas 1989). No public sphere can be brought into existence without "existing industrial-commercial public sphere. . . . Counter public sphere is created in struggle with bourgeois public sphere" (Downey and Fenton 2003); however a struggle transforming the social conditions existing beyond this course of struggle takes aim at creating its own autonomous zones. In this sense, all Occupy movements can be read as the creation practice of both counter and autonomous public spheres. It could be stated that the Gezi Commune also refers to such a process. Even though it was for a short time, an alternative public sphere is created. A new public sphere and citizenship practice are accomplished by the areas occupied and the streets closed by barricades. As alternative and counter public sphere, "play zone" refers to a practice place and time in which the masses can exist freely. This zone is where the communal social relations are created, and makes it possible that the citizens isolated and detached from the social life gather once again: "the [Paris] Commune marks the end of the phantasmagoria of bourgeois modernity by awakening the collective from the dream world of mass consumption and the bourgeois regime and turning the city into a 'play-space' (Spielraum)." Occupy movements proved that the cities can become collectivist and participant by transforming city centers into play zones.

Communal experiences show that "the play-space, typified in his analyis of the commune, aims to construct a new form of a mass public via a new form of collective aesthetic and political practice" (Kang 2014, 193). The transformation of collective aesthetics into daily life practice in the context of both form of action and communal life as well as implementation of a system operates over the practice of participatory, and democratic politics is the sine qua non of the establishment of an alternative counter public sphere. The process of creating an area which refuses social, and political practices, goes even beyond and creates the area kept down by the power, has operated in different forms at different periods of time. Benjamin's view of new public spheres integrates the new forms of solidarity and community into the technological potentials of new media as play zones (Kang 2014, 193). Social media applications exactly constitute the play forms of technology (Hansen 2004, 18) in Benjaminian sense. These play forms have a function of liberating and they are the means of the process of creating liberating play zones. In the process of construction, organization and implementation of a new society into life cooperatively is of crucial importance on the bases of communicational practices. An alternative—counter—public "play sphere" is today realized by Occupy movements and in this process, and it has an important function as being a constituent part of alternative new media. Alternative new media and especially social media in the specific case of the Gezi Commune—in the context of new solidarity, participation, socialization, struggle, and resistance practices—have a central role.

Communism comes back; however, it comes back typically to a historical period of time as always. Communist praxis reveals the ways of revolutionary destruction and construction, as Marx and Engels (1845) stated in the eleventh thesis: "Philosophers have hitherto only *interpreted* the world in various ways; the point is to *change* it." Relatedly "the Idea of communism, itself therefore comes from the idea of practice (from the experience of the real) in the final analysis; it cannot nevertheless be reduced to it. This is because it is the protocol not of the existence but rather of the exposure of a truth in action" (Badiou 2010, 9). In this sense, communist idea can also be described as "revolutionary *aufhebung* (sublation-upheaving)." Badiou notes that communism is a universal tendency in a local process of liberation politics (2012, 13). The locally created experiences embracing the universal show that today's communist political practices both include past experiences and surpass them, as was the case in the Gezi Commune. It is put forward that communism is not a utopia postponed for the future in context of autonomous zones; on the contrary it is an active utopia which can now be experienced. The term "dual power partice" exactly refers to an oppresive oppressive government power against the participatory and collectivist political will of communism that is based on horizontal relations. The experience of Gezi Commune, even though it was for a short time, exhibited that it was possible to build a communalist practice. In this aspect, it was also showed that communism cannot be kept in dilemma of dream and truth, and that the dream may come true.

With the Gezi Resistance, discussions concerning the communalist utopia gained validity again. The Gezi Commune refers to a process in which utopistic thought is awakened beyond a social revolt: "The coming-to-consciousness of a generation is a politically empowering moment, historically unique, in which the new generation, in rebellion against the parental world, may awaken not only itself, but the slumbering, Utopian potential of the epoch" (Buck-Morss 1989, 274–275). Utopia in this context is no longer an unrealizable dream, but it is a realizable communalist life. Communalism is not a problem of the future but it is a problem of now. It can be put into practice in every area that is freed from the government.[11] One of the most distinct examples of this is the long-term experience of the town "Marinaleda" in Spain.[12] As observed in today's social movements, utopia was turned into a social crude fact which can be realized here and now by the communalist activists.

The implementation of utopia inheres in all social revolts. What makes the revolts probable in social sense is social acceptance of possibility of digression, "because the concrete determination of what is utopian proceeds always from a certain stage of existence, it is possible that the utopias of to-day may become the realities of to-morrow" (Mannheim 1979, 183). The action done for turning utopia into reality, beyond refusing existing conditions and transcending them, bears setting up a new life prototype; "Utopia transcends the given social reality; it is not transcendental in a metaphysical sense" (Finley 1967, 6). The main point of utopia praxis is to achieve a creative action of setup beyond destroying the existence creatively. In short, it is the completion of the carnivalesque act of destruction of the original revolutionary thought with utopian setup. Utopia is a desired practice which starts to talk now, "What is utopian is to construct this society out of such hopes, to believe, in other words, that such a society is possible without any other reason or evidence, but that you desire it" (Ollman 2003, 158–159). Temporary autonomous zones are in this sense the reflection of desire now; they are blinking practices showing that this desire is realizable. Ultimately, historical experiences showed us that

> Left cannot renounce utopia; it cannot give up goals that are, for the time being, unattainable, but that impart meaning to social changes . . . Yet why is a utopia a condition of all revolutionary movements? Because much historical experience, more or less buried in the social consciousness, tells us that goals unattainable now will never be reached unless they are articulated when they are still unattainable. It may well be that the impossible at a given moment can become posible only by being stated at a time when it is impossible. . . . The existence of a utopia as a utopia is the necessary prerequisite for its eventually ceasing to be a utopia. (Kolakowski 1969, 70–71)

Utopia requires both creative destruction and construction at the same time. Creative destruction meaning rebellion process is the state in which a carnivalesque chaos is turned into a game. There is a process of rebuilding of place and time participatorily with a collectivist sense in order to make the game everlasting. Utopian creative construction refers to this process. In this sense, utopian praxis of social movements has a carnivalesque prelude. Life has a utopian form in social movements like in carnivals, that is, it is extracted from usual progress. This life gaining a new form is "a topsy-turvy life." It is upside down of the earth (Bakhtin 2001, 238). Creating autonomous zones and building processes of dual power follow the footprints of carnivalesque topsy-turvy being. The moment the society is out of control of the power, it turns life into a "game." Anything suppressed emerges somehow. Transformation of street wars into the games along with that of occupied streets and squares into play zones are the most explicit signs of it. As Bakhtin states (2002, 102), carnival tended to reflect the time game itself, the game both kills and gives birth by re-shaping the old in the new and letting nothing eternalize itself. What is highlighted is the future; utopian features are always contained in rituals and images of the celebrating joy of the public. Rebuilding of the future in social context is possible through seizing the reigning power. For this reason, each carnival can be seen as a rehearsal of seizing the power game. Societies are aware of the fact that it is needed to build a new world in order to rebuild themselves. Therefore, the rebellion is an attempt started to realize the utopia. According to Bercé (2003) rebellion is a very special half-time, a rough form of utopia. Dream and truth are mixed in rebellion, place and time are rearranged, and all the social relations are remolded: "Revolt was a festive interruption of everday life, an exercise of justice and a purifying act that brought renewal" (Bercé 1987, 218). All other Occupy movements and resistances like the Gezi Resistance refer to the processes of creation of a utopian place and time. Occupy movements are not practices blinking for a moment. They are the signs showing that seizing the future game is going on, that the river of rebellion still struggles to outcrop and flow, and the utopian place-time is living underground.

THE GEZI RESISTANCE AND ACTIVIST CITIZEN JOURNALISM

The Gezi Resistance can also be read as a "media war": alternative new media against mainstream media (Lievrouw 2011).[13] In the course of the Gezi Resistance, alternative new media is the autonomous media of autonomous occupy zone. It should also be indicated that it created a state of dual power in media extent. In this sense, it should be noted that the struggle of today's social movements is a simultaneous practice of media. With the help of the Gezi Resistance, the trust of mainstream media was lost thoroughly, and in this period, all events were followed from alternative media. Social media was used effectively, for it provided resistance with important resources: "Remarkably, even as Occupy uses communicative capitalism's networks and screens, its energy comes from a vanguard of disciplined, committed activists undertaking and supporting actions in the streets" (Dean 2012, 216). Social media was used for mobilization of the resistance, communication, and transmission of the news. The frame of news was changed during this course, and news turned into a shared experience from then on. Besides, "in citizen journalism the information is disseminated in bits and pieces as is seen by the people who are witnessing the event . . . the information emerges as a kind of hyperlinked story" (Barnes 2012, 21), the practice of citizen journalism based on social media was rebuilt, recreated by vanguard activists, and it became one of the essential constituents of all Occupy movements.[14] Activists "use Facebook to schedule the protests, Twitter to coordinate, and YouTube to tell the world" (Howard 2011). Social media makes it attainable that new political actors—activists—create themselves in aspect of street movements.[15] Activists created their new political identities over social media, and related to this, they excellently struggled in both real and virtual streets.

Social media enabled the activist to share their thoughts and experiences. It helped dispersed activists take collective decisions, coordinate, and eventually go out and mobilize. As Shirky (2011, 5) stated, social media reduces the costs and time of coordination. Furthermore, it enables information and literature to be distributed. Social media grows awareness over this coordination and distribution. If the "sensibility" of today's social movements is considered, these could be claimed: Social media arouses the spread of sensations, and it serves the generation of a sensational[16] atmosphere, which helps for the massification of social movements, and beyond that, is backed up by the society. Prompting and organization of the sensation in the Gezi Resistance, overcoming the fear of government violence, and boosting courage were possible thanks to social media. In addition to that, it was obvious that there was a sensational and intellectual collectivity alongside a solidarity culture generated among the street protester and the online audience in other words "the Twitter streams helped in maintaining a sense of solidarity between 'physical occupiers' and 'internet occupiers'" (Gerbaudo 2012, 124). Twitter effectively enables the message to flow and helps it be processed in various ways: "On Twitter . . . political news were [*sic*] not only reported on, but also interpreted and actively connected to other topics by the users at the same time" (Maireder and Ausserhofer 2014, 307). Social media, in this context, helped generate an—alternative—network connecting inside and outside. It could be said that this network constituted the biggest power of the Gezi Resistance.

While capitalism transforms all the social life areas, it also transforms the conscious appropriately and recreates it. Journalists, as some of the actors of this process, journalists too have undergone a metamorphosis. Being a producer of urban culture, the journalist reads the city like reading a text and transforms it into a narrative through news texts. Journalists marketing the information by processing and commoditizing it cannot run away from the social relations they produce, and they themselves become commodities bought and sold in the market as well.[17] Commoditization of the information and the journalist signifies the impossibility of a flow of information related to social reality. There are cases in which the transportation of the information about social life and events is impossible. Activists create their own media and make themselves journalists in context of alternative publics far from these meta-relationships.

Activists, who are both activists[18] and journalists individually, obtained the validity of a new journalism form. This form of journalism is based on participation and operated over the network of horizontal relationships which was created thoroughly by the information flow. Citizen journalism showed that there was a possible alternative against the commoditization of journalism: "To me the only difference between a citizen reporter and his professional counterpart is that while one is paid the other one is not. A person who is bound to the boss, not surprisingly, makes the news in accordance with the will of his boss. Citizen reporter is someone whereas a citizen reporter is someone with another job then making news.What frees the citizen reporter at this point is his being outside of these commercial relations" (nabermedya). The Gezi Resistance actually opened up a debate related to redefinition of journalism in the context of existing economy and politics: "Citizen journalism in fact is the only channel pertaining to the definition of journalism. Maybe, the channels defined as alternative media are the unique media which supply the news flow actually . . . Citizenship journalism is going to be one of the most important channels, of the most important topics all over the world, especially in Turkey's forthcoming term." (Can Atalay-Taksim Solidarity). It can be stated that the debate of activist citizen journalism will gain much more importance with rising mass movements.

The Gezi resisters went beyond the silence and blindness of the mainstream media by creating their own media. During this course, social media was used by the activists effectively, and the hegemonic consumerist practice of social media was turned upside down: "The event of Occupy, Occupy's capacity to intervene in our setting, making it and us different from what we were before, has to be understood in terms not of continuity with communicative capitalism but of rupture, of a hole or break" (Dean 2012, 216). The alternative use of new, especially social media, in other words, revolutionary involvement to social communication, changed the paradigms of communication practices. Although there were unfavorable transitions and transformations in the field of media like in each historical time, it revealed that the new media technologies, software, and applications could be used on the basis of alternative political practices. Enzensberger (1970) also mentioned that the homogenization and depoliticization of media did not result from technologies used obligatorily. The foundations of this approach can be found in Benjamin's studies: "What Benjamin really wants to reveal via his analysis of the information industry is the fact that the masses are no longer the passive consumers of the information market or inert spectators of the entertainment industry; rather they are the active producers of information, the aesthetic creators of the new media art, the authours of their own lives in a media space equipped with ever more advanced communication technology, and the consumers of their own lives via social media. They are the new intellecutuals . . . to re-function communication technologies for their own aims" (Kang 2014, 64). It can be argued that there are quite big similarities between the activist citizen journalists and Benjamin's storyteller. It could be asserted that Benjamin's storyteller losing function in the process of transition and transformation has come back with social media, in another form though. Activist citizen journalism will come up for discussion based on the study of "Benjamin and the Media" by Jaeho Kang (2014).

Benjamin notes that everybody can be a media producer by receiving a polytechnic education on media devices and application without any need of special education (Benjamin 1998, 90).[19] With the development of technology, the advance of learning opportunities and user-friendly technologies, although the poverty and deprivation still last, paves the way for the oppressed classes to improve—express—themselves intellectually. Benjamin also notes that the technological developments, the improvements in the technical field, are the bases of the political flourishing of the reproductive writer (1998, 95). It is stated that there are required conditions to enhance a participatory and collective perception in both political and cultural field, in this sense.[20] Benjamin discusses how the transformation of public sphere on the bases of communicative practices transforms the social subjects, and Benjamin "reveals the newly configured conditions of communication and shows how newspapers broaden the horizons of collective cultural and literary practices. The meltdown of the bourgeois literary public sphere also led to the rise of the new media space in which the masses prove to be a multiple social agency as consumers, spectators, the public and even the journalists themselves" (Kang 2014, 58–59). There is an antidemocratic structure which works on vertical relationships by adding all political subjects into production of political ideas through media. Communication technologies reveal that this structure can be replaced by the network of horizontal relationships. In relation to this, the replacement of professional journalism with citizen journalism can be seen as one of the most crucial contributions of technology in context of democratization and participation, "Benjamin's insight into the rise of public writers as an unintended consequence of the development of newspapers is particularly relevant to our understanding of the replacement of conventional newspapers run by trained journalist with the citizen journalists and public newspapers emerging through various forms of Internet communication, such as blogospheres and social media" (Kang 2014, 64). It is observed that the traditional journalist-audience relationship has been disrupted, and a new network of relationships is being created.[21]

The role of the activist citizen journalist is, in fact, precisely put forth by Benjamin. The main point in this approach is to be in the center of the social life and to play the role of one among the perpetrators of the struggle on the basis of protest: "The operative writer's mission is not to report but to fight; not to assume the spectator's role but to intervene actively" (Benjamin 1998, 88). Activist citizen journalist unites the roles of the revolutionist and reporter. S/he is now part of the social protest and the reporter of it. Activist citizen reporters or collectives invite their followers to "gather, understand and create awareness" from the information (Kang 2014, 152). Besides, activist collectives organize online education of activist reportage (Çapul TV, nabermedya) or they show how to use the new technology and applications via the pamphlets and videos of online education.

The activist citizen journalist is both an actor of social protest and a solitary storyteller. In Twitter, followers of those storytellers are both solitary listeners/readers and at the same time collective audience. Activists as participants of social movements and creators of their own media, or as being media themselves, give voice to their experiences by using verbal and visual language, and tell their stories through Twitter. In this course, the storyteller is one of the actors of the story he tells. The Activist citizen journalist is a narrator beyond a protestor. All protestors have ability to become solitary storytellers, and they are collective audience, as well. The definition of collective audience embraces who does not get involved in protests actively but effectively conducts activities on different social networks as being constituent of them. The collective audience is not restricted to the people who only use media tools thanks to the organizations carried out to spread the messages in different forms of media activism in real life besides social media platform. The message is transported individually or collectively to the masses in public transport and in squares. These examples are the signs of the abovementioned (having people watch or listen to the banned tapes in public transport and the displays of the performances showing what was experienced in subways and buses etc.).[22] Then, collective audience includes all the citizens the activists call in daily life beside all the media platforms.

The narratives of social movements today are composed of a sum of fragmented narratives of many solitary storytellers instead of a holistic narrative of traditional alternative media. These fragmented narratives compose a whole in their mobilization. We can see that Twitter has a persistent flow, and that the timeline has a multi-fragmented flow in which many authors as digital producers share their experiences about the ongoing movement in social movement processes. On the other hand, multi-fragmented narratives can present the audience a holistic narrative related to the movement.

There are four features of storytelling according to Benjamin (Kang 2014). It is observed that all these features constituted the essential determinants of the practices of citizen journalists during Occupy movements:

1. As Benjamin also stated, storytelling is one of the earliest communication methods, based on oral tradition. In social movements, the stories have been generated out of experiences, regenerated in each social process. Moreover, they contribute to the perpetual regeneration of social culture.
2. Storytelling is closely relevant to the daily life of society, and it is a collective communication practice in which practical information is shared.
3. Storytelling has distinctive spatio-temporal attributes. These attributes change and transform according to the historical period and the media used. There is continuity between the experiences of the storyteller and the audience.
4. Storytelling rests upon a shared collective experience. The communal attribute of storytelling distinguishes it from other communication models. Experience sharing of the storyteller and audience, being directly in interaction of the collective audience and storyteller, is rooted in the communal structure of this practice.

Activist citizen reporters preserve storytelling over social media channels and applications by rebuilding those four aforementioned attributes as well.

The New types of "storytelling" based on social media created participatory and democratic, in short, communal practice of communication. This practice has also a role in spreading the voice of Occupy movements into the whole society and in gaining support of these movements. What enables the new type of storytelling is the articulation of the real street and digital streets like a Möbius strip. Occupy movements not merely occupied some part of the city, they also occupied the virtual streets. This way, they had the power of carrying the communal spirit beyond physical place and time. That was why they were rather influential. In this course, activist citizen journalists had an important role in the organization of alternative public sphere.

The Gezi Resistance, in the special case of Turkey, is the first massive protest in which the alternative new media is the essential constituent of the resistance. Activist citizen journalists appeared in different resistance areas of Turkey in this course, and they started telling what happened during the resistance, that is, their own stories: "Everyone became a reporter even if it is for a few moments. Everyone has been at some point the primary source of the news. When you have a lot of people reporting, you have a collective account of what is happening. People can follow what is happening via streaming, online, on television, live" (Castells 2012, 114). The very place and time this collective practice of communication emerged from the barricades of the Gezi Resistance and the Commune was Taksim, June 2013. The experience between the storyteller and audience, one of the essential determinants of the Resistance and Commune as a communal practice, is created on social media. A new communication period started when all the protesters went out to become the leading actors of their own stories, and when they started telling their stories themselves. A utopian place and time, in which the producer became the audience, and audience became producer, was created in this period of time. This place and time enabled everybody to express themselves on the basis of collective communication and action.

The manifestation of the activist citizen journalism[23] and its growing importance progressively came together with the loss of reputation of the mainstream media in respect of masses in the beginning of the Gezi Resistance. Especially news channels, the mainstream media ignored the protests. They did not regard them as stories to be told. This brought about the arrival of alternative new media and activist citizen journalists. In a way, activists became media themselves or they created their own media, for the mainstream media became functionless: "People were irritated during Gezi. When they could not see what was going on TV they relied on, they took their phones and went outside, which we had said once and which had remained unreturned" (140 journos). Activist citizen journalists started telling the story of the Gezi Resistance as storytellers by this means. They appeared both as storytellers and characters of their stories. Each activist going out in the Gezi Resistance found ways to express themselves by means of technology in order to reach out to the whole country after a while, and a spontaneous practice of activist citizen journalism revealed itself. An activist with a smartphone became able to practice "citizen" journalism by downloading necessary applications[24]: "No one hears anything unless we write about ourselves and share what we do. I went and bought an iPhone 4s and downloaded all the applications required to make news. I felt I had to make news for I was the one who was there" (@toyotagibiadam). The self-expression process of a society group who could not stand being unseen and unheard was followed by the inversion of the existing approach to media. The social spheres devolved to others (to journalists in this case) were regained and started being used by the society. Under these circumstances, the manifestation of new media was both a need and reaction: "And from May 31 to June 6, the dominant media did not report the resistance anyway, which was not a disappointment for us. The rarely reported news was of course far from the reality. . . . and the resistance had to be reported by resisters, not by others. Çapul TV had already that kind of concept in Gezi Park. Resisters were only in need of a channel in which they came to express themselves and talked about the resistance. The former was a need; the latter was a reaction . . ." (Çapul TV). Whoever could not be informed via the mainstream media or could not express themself started heading for alternative media indispensably: "at that moment, everybody otherized, who could not find a place and information in mainstream, started following Ötekilerin Postası instantly" (Ötekilerin Postası). Since it became a necessity severity of the situation, the activist went through in context of the interaction with their families and circles, and to make news about these: "There are huge numbers of calls on social media, but national channels and newspapers report nothing . . ." My mother called me to learn how it was going. "What's up? You're coughing," she said. "It's because of the gas here" I said. "What's going on there?" she asked me and I replied "Don't you know anything?" She said "No, what's going on?" . . . I devised a strategy, "Mom, share everything I posted on Facebook." "Tell your neighbours not to let the TV manipulate them." "There are massive events here." "Share" I said. There will be no reports on TV in anyway. Her friends use Facebook . . . I thought that the people in Istanbul knew but the people in Anatolia did not know anything. "Take the laptop and show them the videos." That's how I started. "I was shooting videos of 15 or 25 seconds and uploading them on Facebook" (@omurdenden). Citizen journalists endeavored to let their immediate surroundings, and then the whole country, then the world see and hear what was experienced in the Gezi Resistance, what their experiences and stories were. Therefore, they enabled their stories to spread by rebuilding them in different forms.

In the process of mobilization and making news of the Gezi Resistance, social media, specially Twitter, was used quite effectively: "We use Twitter and Facebook accounts of Çapul TV as means of announcement. We heard about everything in streets, and actually, we reached the resisters through this channel. Social media gained profound importance in the course starting from the revolution process in Egypt continuing with June Rebellion. Everything happens there very quickly just with the help of 140 characters. It's very important for us to benefit from it" (Çapul TV). Activists passed on their experiences in different forms to the activists in different places. Therefore, they achieved the development of the collective storytelling spreading all over the country: "Everyone started using their Twitter accounts the day after. This is because we realized that the ones in Istanbul did it. However, Twitter was cramped for space in our life. I worked it out later how critical it was to post a tweet. Besides, a few days later some people came to Ankara from Istanbul. There were people who came from Istanbul to Ankara, broadcasted video-stream and so on. We questioned why we were not doing what they were doing. We realized that there was a need" (@yanniskronos). For the apprehension of social media applications was easy, the activists using different applications became reporters rapidly: "I learned how to use Twitter in a day . . . For example, I'm in Taksim. I write instant news every time. 16:20 this happens. 16:21 that happens. There is not much time gap between the news I report. I always do it by one-minute intervals. I am hyperactive . . . I heard from someone that uploading to Vine is much more rapid . . . I upload every one minute. It's integrated to Twitter, I can share there" (@blackperisi). Activist citizen journalists and their followers learned journalism by making mistakes in the practice in this process. They developed quickly "alternative media ethics" related to journalism in protest practices which were already ethical interventions to life. Citizen journalists tried to give "true news" by using technological opportunities and on the basis of their experiences: "Of course, under the Gezi Resistance there was citizen journalism, there was a process of post and receive news. Naturally, there was serious disinformation among them. We had an advantage in that we knew how to confirm and verify news" (ötekilerinpostası). "When everyone became a citizen journalist, 140 journos had to go one step further, for everyone else was doing that. Then we decided that we were going to collect and verify it" (140 journos). In this course, the ability to make decisions concerning the credibility of news flows over social media was also developed: "On the web, networked publics gain the ability to identify valuable information and call out bad information" (Russell 2011, 147). The skeptic approaches grounded on the approaches of mainstream journalism have some claims that the activist citizen journalism has some problems of objectivity, credibility, validity, and especially ethics. From that point of view, these claims are not valid. It can be said that in spite of all the shortcomings of the activist citizen journalists, they passed a good test for the society in Occupy movements.

CONCLUSION

The Gezi Resistance is the quest of new politics as well as a revolutionary progress. This quest refers to a new process based on the generation and dissemination of the "dual power" state on autonomous zones which will spread all over the living spaces. Gezi changed the paradigm of the revolutionary political struggle, and brought about the refusal of many traditional approaches and memorizations. The Gezi Commune showed that the communist idea is not a utopia postponed for the future, but an active utopia, that it is an "event" with existance, an action to be realized now. Therefore, it signified that it was possible to exceed the limits of the existing world, to realize the communalist utopia with the practices of revolutionary destruction and construction. Online activists and activist citizen journalists in this course played a crucial role in the generation of a revolutionary process that has affected Turkey radically, and in the generation of an active communication network.

As a result, from a more global view, the activists organizing and performing the Occupy protests communicated via social media. Therefore, the new revolutionary politics is about to have all the revolutionary traditions of the past, to perform a new tiger's leap, because the words of Roza Luxemberg "socialism or barbarism" now have a validity unprecedented in history.

NOTES

1. "Referring to 'citizens' media' implies first that a collectivity is enacting its citizenship by actively intervening and transforming the established mediascape; second, that these media are contesting social codes, legitimized identities, and institutionalized social relations; and third, that these communication practices are empowering thwe community involved, to the point where these transformations and changes are possible" (Rodriguez 2001, 20).

2. "Changes in social systems or societies give rise to grievances, social disorientation, or new class or group interests and potentials for collective mobilization. Then there develops a purposive, mass-based movement—coalescing with the aid of ideology and organization—that consciously undertakes to overthrow the existing government and perhaps the entire social order. Finally, the revolutionary movement fights it out with the authorities or dominant class and, if it wins, undertakes to establish its own authority and program" (Skocpol 1988, 15).

3. "The TAZ is like an uprising which does not engage directly with the State, a guerilla operation which liberates an area (of land, of time, of imagination) and then dissolves itself to re-form elsewhere/elsewhen, before the State can crush it. Because the State is concerned primarily with Simulation rather than substance, the TAZ can "occupy" these areas clandestinely and carry on its festal purposes for quite a while in relative peace. Perhaps certain small TAZs have lasted whole lifetimes because

they went unnoticed, like hillbilly enclaves—because they never intersected with the Spectacle, never appeared outside that real life which is invisible to the agents of Simulation" (Hakim Bey 1991, 39–40).

4. "The highly remarkable feature of our revolution is that it has brought about a dual power. This fact must be grasped first and foremost: unless it is understood, we cannot advance. We must know how to supplement and amend old 'formulas', for example, those of Bolshevism, for while they have been found to be correct on the whole, their concrete realisation has turned out to be different. Nobody previously thought, or could have thought, of a dual power . . . What is this dual power? Alongside the Provisional Government, the government of the bourgeoisie, anothergovernment has arisen, so far weak and incipient; but undoubtedly a government that actually exists and is growing—the Soviets of Workers' and Soldiers' Deputies." (Lenin 74, 38).

5. Apart from the example of the "autonomous zones" created by the Zapatistas in the Lacandon Jungle, in India, the Communist Party of India (Maoist), also known as the Naxalites, has set up a government and a number of permanent autonomous zones in the area known as the "Red Corridor" (Myrdal 2012). The recent example of an autonomous zone is "Rojava Communes in Syrian Kurdistan":"In Rojava, to make this system possible, the center of the social system became the commune. The commune, the self-management of the streets, would emerge as the hub of the society" (Knapp 2014). David Graeber states that since 2005 the Kurdistan Workers' Party (PKK) "inspired by the strategy of theZapatista rebels in Chiapas . . . Instead, inspired in part by the vision of social ecologist and anarchist Murray Bookchin, it has adopted the vision of 'libertarian municipalism', calling for Kurds to create free, self-governing communities, based on principles of direct democracy, that would then come together across national borders—that it is hoped would over time become increasingly meaningless. In this way, they proposed, the Kurdish struggle could become a model for a wordwide movement towards genuine democracy, co-operative economy, and the gradual dissolution of the bureaucratic nation-state . . . Councils, assemblies and popular militias have been formed, regime property has been turned over to worker-managed co-operatives—and all despite continual attacks by the extreme rightwing forces of Isis [Islamic State of Iraq ans Syria]"(Graeber 2014).

6. "In our society, which I have concepttualized as a network society, power is multidimensional and is organised around networks programmed in each domain of human activity according to the interests and values of empowered actors. Networks of power exercise their power by influencing the human mind predominantly (but not solely) through multimedia networks of mass communication. Thus, communication networks are decisive sources of power-making" (Castells 2012, 7).

7. "By engaging in the production of mass media messages, and by developing autonomous networks of horizontal communication, citizens of the Information Age become able to invent new programs for their lives with the materials of their suffering, fears, dreams and hopes. They build thir projects by sharing thir experience. They subvert the practice of communication as usual by occupying the medium and creating the message. They overcome the powerlessness of their solitary despair by networking their desire. They fight the powers that be by identifing the networks that are" (Castells 2012, 9).

8. The struggle for autonomy is an issue not confined solely to the real world; the Internet is also a space engaged in a struggle for autonomy. The conversion of the Internet into an alternative information channel and its creation of a digital culture have made it possible not to reject on-street protests but rather to strengthen them. In terms of the network-society, a hybrid counterculture is being created anew through assemblies both in the streets and on digital media. In this sense (on the basis of the assertion that the street and the Web possess a Möbius strip-type of continuous flow) it is possible to speak of a hybrid-public sphere on the basis of the hybrid-network.

9. "The TAZ is thus a perfect tactic for an era in which the State is omnipresent and all-powerful and yet simultaneously riddled with cracks and vacancies. And because the TAZ is a microcosm of that 'anarchist dream' of a free culture, I can think of no better tactic by which to work toward that goal while at the same time experiencing some of its benefits here and now" (Hakim Bey 1991, 40).

10. According to Badiou, Paris Commune is an event, and "an event has, as a maximally true, consequence of its (maximal) intensity of existence, the existence of an inexistent" (Badiou 2010, 222).

11. "Utopian schemas of spatial form typically open up the construction of the political person to critique. They do so by imagining entirely different systems of property rights, living and working arrangements, all manifest as entirely different spatil forms and temporal rhytms. This proposed re-organization (including its social relations, forms of reproductive work, its technologies, its forms of social provision) makes possible a radically different consciousness (of social realtions, gender relations, of the relation to nature, as the case may be) together with the expression of different rights, duties and obligations founded upon collective ways of living" (Harvey 2000, 238)

12. Dan Hancox (2014). *The Village Against the World.* Verso.

13. "Alternative/activist media employ or modify the communication artefacts, practices, and social arrangements of new information and communication technologies to challenge or alter dominant, expected, or accepted ways of doing society, culture and politics" (Lievrouw 2011, 19).

14. "Networked publics are replacing passive consumers and, together with digital tools and news industry economics, changing the way journalism is produced, circulated, and discussed" (Russell 2011, 98).

15. "In OWS (Occupy Wall Street), Twitter, often accessed via smartphone apps,, became a central platform for coordinating protest activisties, for lateral communication among protest participants, but also for sharing news in various formats (text, images, videos) about the protest with a wider online audience" (Deluca and Laeson 2014, 370).

16. "At the individual level, social movements are emotional movements. Insurgency does not start with a program or political strategy . . . But the big bang of a social movement starts with the transformation of emotion into action" (Castells 2012, 13).

17. "In the age of high capitalism, the journalist, who observes the city and reads it as a text, cannot be free from the market and must produce and sell his information as a commodity. Then, finally, he becomes a commodity himself" (Kang 2014, 49).

18. "Social media-equipped activists, who could also be described as 'activist reporters' have become an essential part of the today's media scene. Besides their role as

citizen journalists covering thier own communities and reporting on events that news organizations might mis sor ignore, they also help bridge the gap between news organisations and the non-elite marganalized public, and drive journalists' attention to the voices of ordinary citizens, activists and other marginalized groups" (Kamel 2014, 238).

19. "Aided by the development of technological conditions for 'polytechnic education,' it becomes increasingly possible for the public to become writers, reporters and editors" (Kang 2014, 61).

20. "Participatory media facilitate the access and participation of non-media professionals to the media, and through these participatory media citizens can be active in one of many (micro-) spheres relevant to daily life, organize different forms of deliberation, and exert their rights to communicate" (Carpentier, Brabander and Cammaerts, 2009, 170).

21. "John Hartly . . . believes news audiences have to be encouraged to move beyond what he calls the 'reading public' developed and extended by journalism in past centuries toward a 'writing public' invested in achieving full 'read-write literacy', where everyone is a journalist and everyone has the right and access to the means to create and circulate journalistic information and opinions" (Russell 2011, 140).

22. "Benjamin draws a sharp contrast between the two modes of communication by highlighting the distinctive characteristics of the subjects of communication: the storyteller and the novelist, and the change from the collective audience to the solitary reader" (Kang 2014, 35).

23. "Frequently, video and photographic images of the movements how people recording and transmitting—representing—the movement. Admittedly, some live-streaming actions online present their coverage as journalism. They vary with respect to whether they position themselves as citizen-journalist-activists or as objective reporters covering stories neglected in mainstream media. Others see their tweets, photos, videos, and updates as key components of their activism: through these media contributions they spread awareness of the movement, inform people of ongoing and upcoming actions, and actively increase turnout for meetings and demonstrations. A third, more reflexive aspect of these representational practices appears insofar as the images and reports share in a common name, in a common relation to an occupied Wall Street: this aspect is the self-constitution of the movement as a conscious collective practice. Together, in combination, the multiple streams and images reformat the sense of what is possible through collective action" (Dean 2012, 231–232).

24. "With a smartphone in her pocket, an Occupy activist . . . could become a pan-media outlet, a decentered knot of video, photographs, and blogging that documents and creates and circulates the occupy events" (Deluca and Laeson 2014, 369).

REFERENCES

Badiou, A. 2010. *Communist Hypothesis*. London: Verso.

Badiou, A. 2010. "The idea of communism." In *The Idea of Communism,* ed. C. Douzinas, S. Zizek (eds.), 1–14. London: Verso.

Badiou, A. 2012. "İdeası Komünizm Olan Şeyin Gerçeği de Sosyalizm midir?." In *Komünizm Fikri,* ed. A. Badiou, S. Zizek, 11–21. İstanbul: Metis.

Bakhtin, M. 2002. *Karnavaldan Romana.* İstanbul: Ayrıntı.

Barnes, C. 2012. "Citizen journalism vs. Traditional Journalism." *Caribbean Quarterly* 58 (2/3), 16–27.

Bauman, Z. 1976. *Socialism: The Active Utopia.* London: Allen & Unwin.

Benjamin, W. 1998. *Understanding Brecht.* London: Verso.

Bercé, Y. M. 1987. *Revolt and Revolution in Early Modern Europe: An Essay on the History of Political Violence.* Manchester: Manchester University Press.

Buck-Morss, S. 1989. *The Dialectics of Seeing: Walter Benjamin and the Arcades Project.* Cambridge: MIT Press.

Carpentier, N., L. Brabander, and B. Cammaerts. 2009. "Citizen Journalism and the North Belgian Peace March." In *Citizen Journalism Vol.c1,* ed. E. Thorsen and S. Allan, 163–174. New York: Peter Lang.

Castells, M. 2012. *Networks of Outrage and Hope: Social Movements in the Internet Age.* Cambridge: Polity.

Dean, J. 2012. *Communist Horizon.* London: Verso.

Deluca, K. M., and S. Lawson. 2014. "Occupy Wall Street and social media news sharing after the wake of ınstitutional journalism." In *Citizen Journalism Vol.2,* ed. E. Thorsen and S. Allan, 361–376. New York: Peter Lang.

Downey, J., and N. Fenton. 2003. "New Media, Counter Publicity and the Public Sphere." *New Media Society* 5, 185–202.

Enzensberger, H. M. 1970. "Constituents of a Theory of the Media. *New Left Review* 64, 13–36.

Finley, M. I. 1967. "Utopianism Ancient and Modern." In *The Critical Spirit: Essays in Honor of Herbert Marcuse,* ed. K. H. Wolff and B. Moore, Jr., 3–20. Boston: Beacon Press.

Fraser, N. 1990. "Rethinking the Public Sphere: A Contribution to the Critique of Actually Existing Democracy." *Social Text* 25/26, 56–80.

Gerbaudo, P. 2012. *Tweets and Streets: Social Media and Contemporary Activism.* London: Pluto Press.

Graeber, D. 2014. "Why is the world ignoring the revolutionary Kurds in Syria?." *The Guardian,* October 08. http://www.theguardian.com/commentisfree/2014/oct/08/why-world-ignoring-revolutionary-kurds-syria-isis

Habermas, J. 1989. *The Structural Transformation of the Public Sphere.* Cambridge: MIT Press.

Hakim Bey. 1991. *TAZ: The Temporary Autonomous Zone.* New York: Autonomedia.

Hancox, D. 2014. *The Village Against the World.* London: Verso.

Hansen, M. 1993. "Foreword." In *Public Sphere and Experience.* O. Negt and A. Kluge, ix–xliii. Minneapolis: University of Minnesota Press.

Hansen, M. 2004. "Room for Play: Benjamin's Gamble with Cinema." *October Magazine* 109, 3–45.

Harvey, D. 2000. *Spaces of Hope.* Edinburgh: Edinburgh University Press.

Howard, P. N. 2011. *The Arab Uprising's cascading effects. PSMag,* February 18. http://www.psmag.com/politics/the-cascading-effects-of-the-arab-spring-28575/

Kamel, J. 2014. Reporting a revolution and its aftermath: When activists drive the news coverage. In *Citizen Journalism Vol.2,* ed. E. Thorsen and S. Allan, 227–242. New York: Peter Lang.

Kang, J. 2014. *Benjamin and the Media*. Cambridge: Polity.

Knapp, M. 2014. *Democratic Autonomy in Rojava. New Compass*, October 18. http://new-compass.net/articles/revolution-rojava

Lenin, V. I. 1974. *Collected Works 24*. Moscow: Progress Publishers.

Lievrouw, L. A. 2011. *Alternative and Activist New Media*. Cambridge: Polity Press.

Maireder, A., and J. Ausserhofer. 2014. "Political Discourse on Twitter." In *Twitter and society*, ed. K. Weller, A. Bruns, J. Burgess, M. Mahrt, and C. Puschmann, 305–318. New York: Peter Lang.

Mannheim, K. 1979. *Ideology and Utopia*. London: Routledge & Kegan Paul.

Myrdal, J. 2012. *Red Star Over India: As the Wretched of the Earth are Rising*. Kolkata: Setu Prakashani Publisher.

Negt, O., and A. Kluge. 1993. *Public Sphere and Experience: Toward an Analysis of the Bourgeois and Proletarian Public Sphere*. Minneapolis: University of Minnesota Press.

Ollman,B. 2008. "Why Dialectics? Why Now?." In *Dialectics for the New Century*, ed. B. Ollman and T. B. Smith, 8–27. London: Palgrave.

Purcell, K., L. Rainie, A. Mitchell, T. Rosenstiel, and K. Olmstead. 2010. "Understanding the Participatory News Consumer." *Pew Internet and American Life Project*. http://www.pewinternet.org/2010/03/01/understanding-the-participatory-news-consumer/

Rodriguez, C. 2001. *Fissures in Mediascape: An International Study of Citizens' Media*. Broadway: Hampton Press.

Russell, A. 2011. *Networked: A Contemporary History of News in Transition*. Cambridge: Polity.

Scholl, C. 2012. *Two Sides of a Barricade*. New York: SUNY.

Shirky, C. 2011. "The Political Power of Social Media." *Foreign Affairs* 90 (1), 28–41.

Skocpol, T. 1988. *States and Social Revolutions*. Cambridge: Cambridge University Press.

IN-DEPTH INTERVIEWS

Nabermedya, February 2014
Çapul TV, April 2014
Can Atalay-Taksim Solidarity, April 2014
Ötekilerin postası, April 2014
@140journos, April 2014
@toyotagibiadam, January 2014
@kiyametprojesi, January 2014
@blackperisi, January 2014
@yanniskronos, January 2014
@omurdenden, January 2014

Chapter 6

Social Media and Social Change

The 2011 Campaign for Stopping the Massacre of the Danube Delta Wild Horses

Dana Florentina Nicolae

There is an ongoing Internet joke directly related to the Facebook accounts or pages that deal with issues of a more civil nature such as but not limited to the issue of children and famine in different parts of Africa, or the endless slaughter of dolphins by Japanese fishermen, or any other world occurrence in need of humanitarian support. In a nutshell, the joke refers to the act of liking said pages (pressing "like") by an enormous amount of people and stopping at that, without any consequence or change in real life. The joke stresses the futility of the act and somewhat places a moral burden on those that engage in it, the fact that their undertakings online must coincide with those in reality.

While these pages do reach their aim of getting people informed about the issues and connecting people worldwide to the troubles of their fellow men, a downside developed as a result; not all of the people openly supporting a cause via Internet will actively get involved in the matter, for different reasons.

One of those many reasons is the physical distance between people as people are separated by countries, continents, seas, and so on, and so, the most apt in getting genuinely involved in any act of social change are almost always the people that inhabit the area where the troubles occur. However, even then, there are not all that many people that will participate in acts of protests or of a different nature in order to produce the needed social change. This was and still is the case in many countries around the world and proves that the joke is right.

After the year 2010, after the Arab Spring protests and riots, a change in the general mentality of humankind occurred and started to understand the advantages of promoting elements of social problems online and uniting for a common cause on the same medium as a precursor to more ample acts that aim to repair said problems.

The most important element that precedes these acts of social change is the inherent debate platform that Internet sites offer, where ideas are presented, formed, modified, or rejected, and the end result is always unanimously considered the best course of action given any sociopolitical context worldwide. For example, the Arab uprisings targeted a radical leadership change. In Romania, besides the efforts directed at saving the Danube Delta horses in 2011, in 2013 several protests were organized on Facebook that demanded the halting of the mining project in Rosia Montana because of the cyanide-based technology that could destroy the environment.

The following article will be dealing with one of the cases where people understood the importance of social involvement in an issue related to saving some of the wild horses in the Danube Delta which were unlawfully taken from their natural habitat and almost got sacrificed in a foreign abattoir.

This began per se with a few Facebook pages created shortly before the actual protests and other acts that resulted in the horses' salvation which gave clear information regarding the matter of the horses being transported in a truck. This was provided by a Romanian nongovernmental animal protection organization named Vier Pfoten that had been monitoring the relation between the wild horses and the peasants living in the area.

The Facebook pages received, in a relatively short amount of days, a great number of "likes." Furthermore, 24,000 people signed a Facebook petition that requested the just treatment of the Letea Forest wild horses. Some of the initial comments on the first page created, that summoned people to Letea to make a human chain around the enclosure where the horses were kept, are a clear proof of the people's distress and horror.

Soon after, a spontaneous protest was organized without informing the government about it. While relatively few people participated in it, other people that militated for the cause, among which were some public Romanian figures, were attempting to intercept the truck where some of the horses were taken.

As soon as the truck was intercepted, the supporters for animal rights kept other people informed regarding the situation of the transport via the aforementioned Facebook page. This, of course, drew the attention of the news stations throughout the country, together with the street protests, and resulted in the story being picked up by them and receiving great coverage. Upon the collaboration of the supporters with the Romanian police and Animal Police, the truck was stopped and the horses were released. This happened in late May 2011. Up to this day, similar illegal transports occur, yet they are rarer as the control of horse transportations has become stricter given these events. Suffice to say that in this case, the joke turns out to be wrong.

Naturally the Romanian wild horse issue has ramifications and legal aspects that will be dealt with in the following chapters, but the fact remains that although it is illegal to kill said wild horses without owners in Romania, the sanctions were rarely applied in this case. A great deal of events and social involvement both in the online medium and in real life were necessary to draw attention on the matter and actually have the law enforced.

THE DANUBE DELTA HORSES AND THE PROTESTS

The horses' natural habitat is the Letea Forest in the Danube Delta. In 1938, UNESCO declared the area a world heritage site (UNESCO 2013). Sadly, the forest had been poorly promoted by the Romanian state up until 2011, very few tourist information points existed regarding it, and it is mostly let develop by itself amidst the defectious supervision of the officials caring little for the wild and barely existing species residing there.

There are about 1,000–3,000 wild horses that remain in the area; they are the last ones in Europe and as a result are protected by the local law and by international accords. The problem is that due to the poor involvement of the state in its supervision and development at the time, people that reside in the area simply take some of the wild horses from time to time and declare them theirs without any real legal papers and ultimately sell them to foreign abattoirs for easy money.

One of the many Romanian nongovernmental organizations (NGOs) that deal with animal protection and rights, Vier Pfoten had been keeping a close watch on the situation of the wild horses when, in the spring of 2011, they found out about a horse acquisition in the area, from a local who told them that some of the other locals had sold 54 horses to a foreign company (Four Paws Int. 2012). The horses were supposed to be transported on May 19, and had the approval of the authorities, which was inconceivable given the fact that the NGO stated that there weren't any legal papers that proved that those locals owned all of the horses, a fact which was latter proven to be correct.

In need of help and support, Vier Pfoten created a number of Facebook event pages, "Nu permite Masacrul Cailor Salbatici din Padurea Letea, Tulcea" ("Do not allow the massacre of wild horses from the Letea Forest, Tulcea County!") on May 18 (Anastasia 2011a), followed by the message "Come to Cardon. Support Kuki Barbuceanu (the leader of Vier Pfoten), by going to Letea, they cannot fight the local mafia on their own," "Sustine Campania de salvare a Cailor Salbatici din Padurea Letea," ("Support the campaign for saving the Wild Horses of the Letea Forest") on May 19 asking for donations to help save the horses and "FlashMob—cerem dreptul la Viata al Cailor Salbatici din Padurea Letea" ("FlashMob—we demand the right to life for the Wild Horses in the Letea Forest") on May 20 which urged people to participate at the flashmob set for that date, and wear a white scarf for the horses while stating that "they have the right to a dignified life, they have the right to freedom and we ask for the right to Morality and Dignity, Civilization, in our own country" (Anasta 2011b). They also provided many pictures with the horses being abused, pulled, hit with bats (many pictures showed them bleeding), caught in barbed wire, rounded up, and kept in enclosures which the supporters referred to as "death camps" (Adoptie Distanta 2011).

By May 19 the media took on the subject, and people were waiting for a definitive answer on the horses' future as they were loaded that very morning in the truck and only left on May 20 (Schiahu 2011a), due to the assiduous opposition on the part of the Vier Pfoten members. While the protest/flash mob was in effect, other people were already on the trail of the truck. Prior to chasing the truck, the protesters and members of Vier Pfoten alerted the Animal Police stations in both towns where the truck had passed through. A few of the protesters, among them Andreea Ionescu, kept people updated on the Facebook page about the location of the horse truck. The posts she made gave detailed information about the truck and where it had been stopped:

> ATTENTION ROMANIA . . . EVERYBODY FROM ROMANIA . . . .STOP THIS TRUCK . . . Masina are nr. BZ 13 SIR (The car has the following number BZ 13 SIR) . . . leaves Tulcea now . . . on its way to SF.GHEORGHE Slaughter house . . . with 49 horses . . . .19 have infectious anemia . . . lots of colts . . . ROMANIA . . . STOP THIS TRUCK. . . . BZ 13 SIR . . . (Voica 2011).
>
> She then continued: 'Right now, the Letea horse truck is being pulled over by the Slobozia town police, in Viziru, they are checking for the papers, we still don't know the abbatoir where they'll be slaughtered. There are 9 cars tailing the truck, they have been behind the truck since (they were in) Tulcea County, around 40 people, (they) are not enough to stop the truck from entering the abattoir with the horses, there is no more press. Send this message to the press, that 40 people are tailing the Letea horse truck, so that they can make a human shield at the abattoir entrance!' ('Acum camionul cu caii din Letea este oprit de politia Slobozia la Viziiru, se verifica actele, in continuare nu se stie abatorul unde vor fi macelariti. Sunt 9 masini care au plecat dupa camion, sunt masinile care merg in spatele camionului din Tulcea, in jur de 40 de oameni, insuficienti pentru a opri intrarea camionului cu cai in abator, presa nu mai este . Trimiteti la presa mesajul ca 40 de oameni insotesc camionul cu caii din Letea, sa faca un scut uman la intrarea in abator!'). And finally, the last post: 'The truck is on the way to Buzay. Definitely the slaughter house is Casalco Sfantu Gheorghe' (Voica 2011).

After the flashmob in Victoria Square, Bucharest (Stirileprotv.ro 2011), many people proceeded to go to the abattoir where the horses would presumably be brought in order to keep the employees from taking them in. However, they were to find out that the truck had been stopped by the authorities in a town in Braila County. On the night between May 20 and 21, after 300 km and approximately a day on the road, the horses were released from the truck, after sitting outside the police 'station for four hours. Many were famished and thirsty, and one of the horses had died. They were housed at a farm in Braila County (Vier Photen gallery 2013), in the town of Urleasca and for the next few days people came to bring them food and to guard them. Given their abuse and wounds the horses also received medical care although, unfortunately, a few died. Their end was also ensured by the poor conditions during the transport.

Some of the locals came to claim some of the horses that were saved, and had trouble doing so as by that time, the Animal Police investigations revealed that the papers used for the transport were shady and illegal, so those horses remained at the Urleasca farm for a while, together with free ones. The truck driver was charged with bad treatment and cruelty towards animals (Law Nr. 205, year 2004), and it became clear that it was anything but a safe and proper animal transport.

The locals that declared themselves the owners of the horses, together with the local authorities in Tulcea County, said time and time again that the papers were legal and that Vier Pfoten had blown everything out of proportion (Sultanoiu 2011), and that they would eventually sue the NGO for all of the lies it propagated about the status of the horses. On top of all this they said that they would send another horse transport to an abattoir as soon as possible, but that never came to pass, as an emergency state decision forbade any other horse transports coming from the area.

It was only after three months that the money necessary for returning the horses to their natural habitat was raised, and soon after, they were taken back to the Danube Delta (Green Report 2011). Unfortunately, a couple of locals got into arguments with the Vier Pfoten members before any of the horses could be released. They claimed that the horses were theirs and that they were unjustly taken from them without being paid anything (Observatortv 2011). But the situation was explained to them, and they realized that they had to accept the placement of horses in quarantine before they were released or returned to their owners, if they had the means to prove so, in the absence of microchips implanted in the horses. There were also enough funds to transport fodder along with them so they could have a richer source of sustenance. Prior to these events, only Vier Pfoten had bought fodder for the horses, out of donations.

Given that the wild horses come from a natural environment and not a man-made one, another issue related to them that was again overlooked by the local authorities upon their sale was the fact that many suffer from different diseases, and in the case of the horses in the truck many were afflicted by infectious anemia, a fact that did make them suitable, by law, to be sent to an abattoir, as the disease is not contagious to people (Vanghele 2011), although it does legally require a special note in which the buyer declares that he/she accepts the horses with the disease. However, the protesters stressed that while some horses did have anemia, others did not and consequently, there was no legal ground for them to be sent to the abattoir together and put in conditions in which it was easy for them to catch the disease.

Upon conducting the blood work, the doctors concluded that from a total of 54, only 19 of them suffered from infectious anemia (Mediafax 2011), and as a result it was illegal to transport the ill ones together with the healthy ones to an abattoir, due to the possibility of the latter getting infected; and on top of all this, 33 of the horses weren't microchipped, which meant that they had no legal owner.

To this the authorities in Tulcea County, together with the locals, declared (Marian Sultanoiu 2011) that there is no such thing as a "wild horse" species in the region, that the horses were owned by the families that had settled there, and that they were merely raised in an unorthodox way, by the people, by letting them "run wild," and that is also why actual ownership is so hard to be established. Of course, the horses multiplied freely thus creating these special caballine communities.

They went on to say that the horses constitute somewhat of a problem for the Danube Delta ecosystem because they continue to grow in number and to feed chaotically with the resources at hand, thus endangering the natural reserve at Letea Forest. Not only that, but the Tulcea authorities received numerous complaints from locals and associations asking it to solve the problem of the damages made by the horses in the area.

These statements were attacked by representatives of Vier Pfoten (Sultanoiu 2011) who insisted on the fact that while the overgrazing issues are in effect, the horses are an integral part of the Danube Delta's ecosystem, a healthy presence for the development of the area and should be protected at all costs—rather, they should be helped along the way with fodder so as to eliminate the problem altogether without harming or endangering the horses—and they have repeatedly asked for an accord with the Romanian government in this sense.

THE AFTERMATH AND THE VODAFONE COMMERCIAL

Luckily, together with the horses being saved, and the intense media coverage, a set of actions were conducted by state officials in this sense, and from then on, a harsher checking system was implemented by the government, in order to discourage such acts that border on poaching. Furthermore, the accord for the protection of the wild Letea horses became a fact approximately three months later, on August 28.

Up until then, the NGO had to deal with the numerous problems caused by the supposed owners of horses and the local authorities that backed them up. One representative of the Animal Health department in Tulcea County, Marinela Tertis (Romanialibera.ro 2011a), who refused to sign the accord, declared that the signing of the protocol was blocked because Vier Pfoten wanted to take actions that are illegal, such as the prohibition of selling horses to abattoirs by their owners. Not only that, but keeping the ill horses in an uncontaminated area in Urleasca is also illegal.

The accord or protocol (Protocol de Colaborare), entitled "Caii Salbatici—Letea" ("Wild Horses—Letea"), clearly states the purpose of the accord which targets the decrease of the phenomena that resulted from the uncontrolled reproduction of the wild and semi-wild animals by implementing a program of animal reproduction and medical control through the identification, vaccination, and sterilization of the horses without an owner. It featured the following issues: that the signing parties, in this case (initially the accord was to be signed by Vier Pfoten, several local institutions from Tulcea County such as the National Environement Guard, the office for animal reproduction, the frontier police, local councils and prefecture, Regia Nationala a Padurilor [ROMSILVA] and the administration of the reservation of the Danube Delta Biosphere, but most of the administrations in Tulcea county declined.), Vier Pfoten, the Ministry of Tourism, the local council of C.A. Rossetti (the district where the horses were rounded up), and the administration of the reservation of the Danube Delta Biosphere, agreed to make an impact study of the wild horses' influence over the Letea Forest area and its surroundings; the development of a strategy for managing invasive species and preserving the biological diversity of the Danube Delta; the halting of the vegetation decline by providing the horses with fodder during the winter months; veterinary check-ups for the horses; the identification of the horses and their owners and creating a data base with these elements; the implementation of educational programs for the local horse owners in the area regarding the conditions of raising and owning animals and the boosting up of touristic interest given the natural potential of the area, as Letea is the only place in Europe where people can find wild horses; and the creation of social opportunities for the local communities by involving them in the project.

These measures were to be implemented in three steps. The first step dealt with the promotion of the project, through flyers and announcements related to the project, informing the local animal owners about the laws regarding animal health regulations and finally informing the local authorities about the project being in effect.

The second step dealt with building enclosures for the anemic horses in the area, establishing the proper size for them, and finding economic agents interested in buying them (the ill horses).

The third step dealt with catching the horses and identifying them; taking blood samples and microchipping them; creating a database; establishing if they had owners or not, and if they did, informing the owner about their legal obligations; the segregation of ill animals, selling them or buying them from the owners, disposing of the cadavers, and establishing an animal cemetery.

However, since most of the initial desired partners refused to sign the accord (Romanialibera.ro 2011b), and the Ministry of Tourism took notice of it, the accord had several elements added to the aforementioned steps that involved the exploitation of the tourism potential of the horses, and was signed on August 28 (Agerpres 2011), at Letea.

Then Minister of Tourism, Elena Udrea, declared in a conference held at Cardon, on the same day that: "We must use the presence of the horses in Letea to draw more tourists to the area. We believe that this accord that we've named 'The Letea Protocol' marks a new beginning for the horses that have suffered up until now, and also for the locals" (Agerpres 2011).

She continued: "There is a story of the horses in the Letea Forest and we must take advantage of this desire people have to visit the area. That is when tourism will flourish in the area, residential hotels, restaurants and other facilities will be built, that will, in turn, create jobs" (Elena Udrea, August 28).

As a result, a decision was made by state authorities to surround the area (about 1,000–2,000 accres) where the horses reside with a fence so they could live as they have until now and to also make it easier for them to be supervised and visited by tourists. Furthermore, the ministry promised to take care of everything related to the feeding points for horses and building information points in the area for tourists, and establishing visitation routes.

Kuki Barbuceanu, the leader of Vier Pfoten confimed during the same conference that during the intense media attention given to the salvation of the horses, the number of tourists in the area had increased three times the average (Kuki Barbuceanu, August 28, 2011).

Undoubtedly, these events and their technological aspects had a great impact on the Romanian society and were consequently picked up by Vodafone, a phone company, in one of their many region-specific commercials, approximately a year after.

The commercial film entitled "50% mai multa putere sa te implici" ("50% more power to get involved") (Vodafonebuzz 2012) features an inspiring message about participating in acts related to greater ideals and social change and stresses the importance of the many different communication devices that can bring people together for the same purpose, and lists the Letea horses' rescue as the proof.

Its script follows the events of the rescue by heart, it begins with shots of people protesting in a square for the horses, continues with images of people getting informed about the "Letea horse massacre" either through online news or by using their mobile phones, and ultimately these images were superimposed with different messages that were made to look like the initial reactions about the horses' ordeal.

The next shot features a truck (carrying horses) and someone driving towards it while being superimposed with one of those messages that urge people to go to Letea and take action. Furthermore, the protesters from the beginning walk on the road toward the truck and the latter stops right in front of them. The doors fall down and the horses run wild.

However, this did not sit well with some of the real protesters as they begun to criticize the audacity of the phone company's claims of being among those that helped save the horses. The real protesters also criticized the false depiction in the commercial of the events as they occurred, starting with the fake superimposed Twitter messages, and ending with the large number of protesters that stopped the truck unescorted by the police (Vodafonebuzz 2012).

However, to be frank, based on the commercial, there is no mention of Vodafone taking credit for the horses being saved, rather it makes a statement about the importance of being connected constantly (via mobile phones, tablets, laptops, etc.) to the main events in the world, the events that influence one directly, and of course, Vodafone provides such means for people. The company asked Oana Pellea, a Romanian actress and public figure who was actively involved in the horses' rescue, to be the voice of the advertising campaign, which she gladly accepted, after publically thanking Facebook for existing (Schiahu 2011b) and for its impact in 2011.

Instead of a conclusion, I would like to present a reflection on this type of protests. As they are increasingly gathering people (see also the "Rosia Montana" protests in 2013), they are being assimilated in a vision about the world that betrays an impossibility of young people to understand the political elite's actions, and the only attempts at change via protests are motivated by the "better marketed ideals" on Facebook or other social media sites. This betrays the lack of a political culture in young Romanians (because most of the people participating in protests are young adults).

Although there is a history of manifestations in large numbers—the Romanian communists encouraged these rallies, albeit for political propaganda purposes, after the 1989 Revolution conducted by people gathered to radically change the regime—there were few protests in the years that followed, despite people being generally displeased with the state of things and the way the country is run. It was only in 2011 that Romanians began to understand the importance of uniting for a common good and the power and visibility that large numbers of people have.

Naturally all of this happened due to the rise of social media. There is a noticeable shift, due to recent events, from state media to social media rendering the former less dominating. There is a generalized trusting attitude towards social media-expressed opinions rather than information from the state media.

In fact, it appears to be the place of genuine involvement due to the fact that the people that use social media or have accounts on these sites and interact with others have their real name and provide real information about themselves, and as such they account for their words, reactions, and opinions. Sadly, these social media products are only available to a certain amount of people who have access to the Internet and are inclined to engage themselves in social site activity.

Perhaps these protests, either solely on Facebook where people involve themselves at a purely declarative basis, or the full Facebook protest combined with street protest, herald a new political conscience in the minds of Romanians that is just being formed after years of being confused about the state of things and having only got their information from the state media.

REFERENCES

Adevarul. August 28, 2011. "Caii de la Letea se întorc în mediul natural." adevarul.ro/locale/tulcea/caii-letea-intorc-mediul-natural-1_50abbd967c42d5a6637f8769/index.html.

Adoptie Distanta. May 30, 2011. "They were born to be FREE" https://www.facebook.com/media/set/?set=a.219057628112583.62570.115974898420857

Agerpres. August 29, 2011. "MDRT şi Vier Pfoten au semnat ‹Protocolul de la Letea›." http://www.agerpres.ro/english/index.php/news-of-the-day/item/79734-MDRT-si-Vier-Pfoten-au-semnat-Protocolul-de-la-Letea.html

Anastasia, Mirela. May 18, 2011a. "Nu permite Masacrul Cailor Salbatici din Padurea Letea, Tulcea." https://www.facebook.com/events/210379065659681/.

Anastasia, Mirela. May 19, 2011b. "FlashMob-cerem dreptul la Viata al Cailor Salbatici din Padurea Letea." https://www.facebook.com/events/166383193422194/

Four Paws International. February 14, 2012. "Caii Salbatici din Letea salvati de VIER PFOTEN - povestea in imagini." http://www.youtube.com/watch?v=vUS9qkyEEkY.

Green Report. May 28, 2011. "Focarul de anemie infectioasa blocheaza transportul cailor de la Letea." http://www.green-report.ro/focarul-de-anemie-infectioasa-blocheaza-transportul-cailor-de-la-letea/

Letea. https://www.facebook.com/events/116195871797884/.

Mediafax. May 24, 2011. "Cazul Letea: S-a terminat recoltarea de sânge de la 52 de cai; 33 nu au microcipuri." http://www.gandul.info/stiri/cazul-letea-s-a-terminat-recoltarea-de-sange-de-la-52-de-cai-33-nu-au-microcipuri-8282768

Observatortv. August 28, 2011. "Caii de la Letea sunt salbatici, dar cu proprietari. . ." http://observator.tv/social/caii-de-la-letea-sunt-salbatici-dar-cu-proprietari-41365.html

Protocol de Colaborare. 2013. "Caii Salbatici—Letea" (Collaboration Protocol. "Wild Horses—Letea"). http://storage0.dms.mpinteractiv.ro/media/401/521/7568/8528499/11/document-nr-4-proiect-protocol-colaborare-3.doc.
Romanialibera.ro. June 1, 2011a. "Scandalul cailor din Deltă continuă: Autorităţile refuză să semneze protocolul pentru protejarea lor." http://www.realitatea.net/scandalul-cailor-din-delta-continua-autoritatile-refuza-sa-semneze-protocolul-pentru-protejarea-lor_840846.html#ixzz2fz7R1K2f
Romanialibera.ro. June 2, 2011b. "Scandalul cailor din Deltă continuă: Autorităţile refuză să semneze protocolul pentru protejarea lor." http://www.realitatea.net/scandalul-cailor-din-delta-continua-autoritatile-refuza-sa-semneze-protocolul-pentru-protejarea-lor_840846.html#ixzz2ftD3gQgg
Stirileprotv.ro. May 20, 2011. "Petitie pe Facebook: 24.000 de oameni vor sa ii salveze pe caii din Letea." http://stirileprotv.ro/stiri/social/petitie-pe-facebook-24-000-de-oameni-vor-sa-ii-salveze-pe-caii-din-letea.html
Schiahu, Mihai. May 21, 2011a. "CAII DIN LETEA AU FOST SALVAŢI: Tocmai ELIBERĂM CAII. AU FOST SALVAŢI DE ABATOR, cel puţin pentru moment." http://www.gandul.info/stiri/caii-din-letea-au-fost-salvati-tocmai-eliberam-caii-au-fost-salvati-de-abator-cel-putin-pentru-moment-8277460
Schiahu, Mihai. May 21, 2011b. "OANA PELLEA, despre salvarea CAILOR de la LETEA: Toate mulţumirile trebuie adresate acestei PĂLĂRII NUMITE FACEBOOK." http://www.gandul.info/stiri/oana-pellea-despre-salvarea-cailor-de-la-letea-toate-multumirile-trebuie-adresate-acestei-palarii-numite-facebook-8277476
Sultanoiu, Marian. May 24, 2011. "SCANDALUL CAILOR DE LA LETEA. Sătenii se plâng că le-a fost încălcat dreptul de proprietate. Şeful Vier Pfoten: asta este chiar culmea tupeului şi a ipocriziei la ţărani." http://www.gandul.info/stiri/scandalul-cailor-de-la-letea-satenii-se-plang-ca-le-a-fost-incalcat-dreptul-de-proprietate-seful-vier-pfoten-asta-este-chiar-culmea-tupeului-si-a-ipocriziei-la-tarani-8280932
UNESCO. 2013. http://whc.unesco.org/en/list/588.
Vanghele, Ovidiu. May 20, 2011. "Cazul Letea: 49 de cai, 19 având anemie infecţioasă, duşi la abator. Carnea va fi exportată." http://www.mediafax.ro/social/cazul-letea-49-de-cai-19-avand-anemie-infectioasa-dusi-la-abator-carnea-va-fi-exportata-8276956
Vıer Photen Gallery. May 2, 2013. "Caii din Letea au ajuns la Urleasca." https://www.facebook.com/media/set/?set=a.10150189512631644.305541.302068656643&type=3
Vodafonebuzz. October 1, 2012. "50% mai multa putere sa te implici—Vodafone" http://www.youtube.com/watch?v=Uejfh8qyQsY
Voica, Steluta. May 21, 2011. "Salvată de la sinucidere pe Facebook, salvează caii de la Letea tot pe Facebook." http://www.realitatea.net/salvata-de-la-sinucidere-pe-facebook-salveaza-caii-de-la-letea-tot-pe-facebook_837122.html#ixzz2ftbfVVI6.

Chapter 7

Active or Passive Citizen

The Influence of New Media Over Local Communities' Participation

Ilona Biernacka-Ligieza

Participation is one of the most important elements of democracy and is closely associated with a number of processes that link with communication and the access to social decision-making processes. These processes to a greater or lesser extent use information and communication technology (ICT) and the Internet as an aid, even though they themselves are not dependent on the Internet. This article looks at the ways the affordances of the Internet and Web 2.0 services apply (or not) on the level of local democracy. Globalization does not render locality irrelevant, but it challenges it as was emphasized for instance by the concepts of glocalization (Featherstone, Lash & Robertson 1995, 25–44), hyperlocality (any reference), or translocality (Appadurai 1996). In general, these concepts suggest that encounters of the local and the global can be productive in terms of modifying locality by placing it in the wide context of the global and vice versa. In the case of using digital technologies in a local context, we are interested in the width but in the depth. Looking at the implementation of the digital technologies into the life of the local community, we are not concerned with the question not how the local gets modified by the globality but if it can get intensified. Some give skeptical accounts of the decaying sense of participation in the life of the local communities and argue that in the late modern period real (offline) local communities dissolve into isolated individuals (Putnam 2001). Potential intensification of the local participatory energy on and through the Internet is therefore more important regarding the fact that it is missing even in the real life of local communities.

THEORETICAL FRAMEWORK: CONCEPT OF LOCALITY IN THE GLOBALIZED WORLD

The impact of globalization on local media and society within specific cultural and national contexts could, thus, be multifaceted. Many scholars (Hegde 2005; Thussu 1998, 2007; Hafez 2007) analyzing today's media trends notice that the global media concerns if they want to achieve the market must be aware of local needs, and they must adjust their programs to the local context. Moreover it is seen that locality year by year is perceived as one of the most important factors of media economy. It could be seen as the process of mutual cooperation on the local, the national, and the global level. The process of strengthening both the local and the global media can be seen in more countries each year. This tendency allows media scholars to hope that globalization, characterized by multiplying audience, does not necessarily lead to cultural homogenization but, rather, reinvigorates cultural diversity in new ways.

Current trends such as technological convergence, free movement of goods, services and capital, as well as market mergers and acquisitions, often jeopardize the potential growth of local cultures, particularly in developing countries and in transition countries. Cultural industries are increasingly concentrated in the hands of a few players who have advanced technical capacities for producing and distributing creative products.

On the other hand, in order to flourish, what local cultures need is not to be preserved unchanged, like insects in amber, but to live and develop in their own way, as they themselves choose. And in this, they can benefit from the use of the new tools of information, knowledge, and communication which they can obtain only by increasing their connections to the larger world. Cultural exchange is also a key factor in breaking down the barriers to good relations between peoples and human groups and a powerful instrument of "preventive diplomacy."

Bauman in his deliberations on globalization has devoted a lot of attention to locality and its retardation in the contemporary world. We no longer need a transmission of a material carrier to transmit information, so Bauman decided that "a town and local people have little to do with local community" (Bauman 1998: 31). Quoting Lazarfeld, he concluded that in the globalized world there exists no "local agora," no place for the opinions of local authority; a place in which "plans of actions are developed, private matters become public, opinions are created, courts are organized and punishments given" (Bauman 1998, 31). It seems, however, that these local debate forums still exist, and, what is more, they create new spaces for public debate.

The process by which, in the globalized world, individual cultures relentlessly tend to decrease in importance or to be overwhelmed by giving room to a single global culture, is often source of concern. The resulting bewilderment materializes in one of two ways: exaggerated devotion to the past, often a reinvented past, in the name of tradition and the opposite tendency, namely the noncritical adoption of any cultural input coming from the outside.

Arjun Appadurai describes the production of locality as the object of ethnography. For him, "locality is an inherently fragile social achievement," that must be maintained carefully against various kinds of odds, especially through ritual. Not only boundary maintenance, or foundational ritual, but also rites of passage produce locality as both a spatial construction and as a way of embodiment—through the production of local subjects. The glocal approach suggests that local cultures may be valorized and revitalized through the means and opportunities offered by globalization in order to make them a source of enrichment for all the individuals involved. Instead of a superficial interest for local habits and folklore, glocalization involves the belief in an active multiculturalism based on the acknowledgement and valorization of differences between cultures, and on the promotion of cultural exchanges whereby people from diverse societies can express and communicate to each other their visions, values, and traditions. Robertson ultimately hopes to dispel the "notion of homogenization of all cultures" by the forces of transnational capital (Robertson 1995, 25). Glocalization is not what Thussu (1999) refers to as "desi globalization—when foreign business interests and media emphasize their native credentials to present a more acceptable face of globalization" (Thussu 1998, 46). Instead, Luke argues "glocalization attests that every global, virtualizing force is met by a stubborn alterity, a 'dialectical intensification of both globalism and localism' " (Luke 1997, 89). Concerning the implementation of new communication technologies structure into the local public sphere we must take into consideration not only challenges but also doubts and threats connected with global technology development, and that is one of this paper's main issues.

AFFORDANCES OF THE NEW MEDIA AND DEMOCRACY

Nowadays, the one-way media, which are characteristic of mass society and include the press, radio, and television, are losing with interactive, digital means of communication. New ICT may serve not only for work or entertainment, but also for building up the society. We associate with ICT such hopes as for example, to expand the opportunities for public information about the intentions of the government or the increased participation of citizens in co-decision-making about the shape of the state, which is particularly important in the face of declining interest to participate in public life. The impact of ICTs on political processes and state-citizen relations is determined by means of such concepts as teledemocracy, digital democracy, cyberdemocracy, e-democracy. Regardless of the name and scope of the concept, common to these ideas is the belief that various features of new technologies—interactivity, faster mode of information, communication feedback—may, in a positive way, influence on democratic mechanisms (Białobłocki, Moroz, and Nowina-Konopka 2004; Chadwick 2006; Zissis and Papadopoulou 2009; Nowina-Konopka 2009; Cindio and Schuler 2012).

New communication technologies are being used by local communities mainly in two forms as e-government (top-down process) and e-participation (bottom-up process).

There is no doubt that ICT is the strong support for local administration. Proper and smoothly functioning e-government system seems to be standard for each democratic country.

But together with changing the local administration standard also due to spread of Web 2.0 media[1] the perception of representation is being reconstructed. Web 2.0 itself may not strongly affect citizens' attitudes toward community involvement and participation, but can be used to create conditions for a more involving, inclusive, and transparent democratic culture.

Both mentioned activities (top-down and bottom-up) are two halves of e-democracy. As we see new communication technologies cause many changes in contemporary local public spheres. These may be changes in how the internal organization of the public sector works, that is, the technology can be used to improve collaboration and integration between different levels of government and between offices. The second type is amended to increase the availability of online information about the work of public institutions and the fact that the public sector will try to arrange its activities for the benefit of the people so that the center of activity is shifted towards self-service (Grönlund and Anderson 2007; Yates, Gulati, and Tawileh 2010; Nam 2010). According to Trechsel:

> e-democracy consists of all electronic means of communication that enable/empower citizens in their efforts to hold rulers/politicians accountable for their actions in the public realm. Depending on the aspect of democracy being promoted, e-democracy can employ different techniques for: 1. increasing the transparency of the political process; 2. enhancing the direct involvement and participation of citizens; 3. improving the quality of opinion formation by opening new spaces of information and deliberation (Trechsel and Mendez 2005, 5).

E-democracy relates to those uses of ICTs which provide either novel or more efficient, practicable means for citizens to exercise influence in the governing process. In other words, this term applies when ICTs are used to revolutionize the relationship between citizen and representative.

E-government, however, is a term that can be invoked to describe those uses of ICTs which do one of two things. First, under the paradigm of organizational theory, they may make the interaction between different agencies and different bureaucrats more efficient, as they do for private firms. Second, they may make the relationship between the government bureaucracy and the citizen, in the sense of a political consumer, more comfortable and efficient by reducing transaction costs. To put it another way, when a citizen can pay a fine or fill out a property tax form online, this is an example of e-government. In many cases, e-government and e-democracy initiatives may emerge at the same time as various levels of government increase their reliance upon, as well as familiarity with ICT's. Regardless of who takes the initiative, action may be a unidirectional process (one-way) or bidirectional (two-way). An example of the first of these is the dissemination of information; of the second, consultation on draft legislation. For several years, governments throughout the world have been seeking to provide electronic access to government services. Key reasons for this public sector reform have been to increase the efficiency of government operations, strengthen democracy, enhance transparency, and provide better and more versatile services to citizens and businesses (Coe et al. 2001; Ho 2002; Giritli Nygren 2009; Ifinedo and Singh 2011).

According to Macintosh, e-participation means the use of ICTs to broaden and deepen political participation by enabling citizens to connect with one another and with their elected representatives. The emphasis of this definition is the involvement of all stakeholders in democratic participatory decision-making, instead of just the top-down initiatives of the governments (Macintosh 2008). Romsdahl understands e-participation as one of the many methods of public deliberation. Public deliberation is not a new concept in itself, but e-participation holds potential to revitalize it (Romsdahl 2005, 44). On the other hand, some scholars approach the concept of e-participation from a critical point of view (Fountain 2001).

SPECIFICITIES OF THE NEW MEDIA USE IN THE LOCAL CONTEXT

Internet is well known for its suggestive spatial dimension and strong, though immaterial, geography consisting of areas that are experienced as local. The capacity to motivate people's concentrating on specific shared "places" was one of the earliest noticed phenomena described as creation of virtual communities (Rheingold 1993). Geographical proximity is not required. It is an "imagined community" or a "community without proximity," consisting of people who have never met but enter into more intimate relations in the virtual world, than in reality with the people living in the same town or house (Barney 2006).

It is worth noting that along with the only virtual communities, there are still local communities that exist both online and offline despite having the required tools and possibilities to create virtual community. There are also only real communities which do not have access to digital tools because they either do not feel the need to use them, or they are "digitally excluded" and have no access to the technology. It is worth noting that in a way, traditional local media also integrate real-life communities because most people living in a given region or town perceive themselves as a part of the community only in their minds and rarely enter into direct interaction with each other (Anderson 2006). Urry, deliberating on the community mobility in "network society," noticed that Bell and Newby (1976) have identified three useful categories concerning the meaning of the community: proximity, locality, and community. These categories are also useful in the discussion of the local media, because one can identify different kinds of media used for communication by a given community.

Internet and its Web 2.0 services are used by the many as a space for debate, intervention, and grassroots journalism, which are the activities that may have also local-specific functions.

The first opportunity which the Internet creates is the chance to create a local debate forum. It is a community of people who live in the same place, go to the same institutions to take care of official matters, people to whom apply the same rules set by the authorities, people who do the shopping in the same places and so on. Another potential of the Internet is a possibility to fill the gap in the traditional media market. Where there is a lack of "official" information, and there are too few journalists in remote places, it is possible to create a communicative platform on the Internet which would be based on bottom-up initiative coming from the citizens.

The capacity of new media to increase the quality of democracy has been the matter of conflict between the skeptics and the utopians ever since the debate over the digital democracy commenced, and the same goes for the Internet in the local context. Tests based on involving local communities in political debates and decision-making via ICT at the beginning of the twenty-first century showed diversified interest from the citizens' side and the researcher had different opinions about the digital public sphere at the local level (Wilhelm 2000, 2004; Hoff, Horrocks, and Tops 2000; Norris 2001). The first experiences of virtual commune or city were driven from those countries which were familiar with highly advanced communication technologies (e.g., the United States [California], the Netherlands [Amsterdam], Germany [Berlin], Denmark [Jutland], Sweden [Gothenburg]). Nowadays we can see dozens of initiatives being taken on that field in different countries, both in countries with highly developed ICT and long democratic tradition (e.g., Norway [Modalen, Agder, Larvik, Molde]; Finland [Tampere]) and those with reborn democracy and average ICT development (e.g., Poland [Małopolska, Lower Silesia]; the Czech Republic [Prague, Brno]). With this rapidly growing transformation of the local governance structure, the researchers show a lower level of skepticism about ICT being used as one of the main tools of that transformation than it had been at the beginning of that process (Orihuela and Obi 2007; Jeitziner 2009; Meier 2009). Analyzing the influence of the ICT on the democracy we are able to see that new communication technologies may increase access to political decision-makers and, in a broader sense, contribute to public opinion formation both on global as well as local issues. The key element is that the Internet has a low threshold: ICT makes it much easier for "ordinary people" to get through with their messages than any other medium.

Local government has a critical role to play in this e-democracy time as the branch of formal democracy that sits closest to real people and real communities, that deals with people's local concerns, and that helps people shape the places in which they live. Most local governments in Western democratic societies put high impact on implementing the Web 2.0 agenda, but they are also conscious that the real import of Web 2.0 is not in the end about the use of online tools, but rather in the extension of the Web's collaborative, participative logic to all of the council's activities. But when we compare Eastern democracies (e.g., Poland) we see local government scarcely using Web 2.0 technologies and in a very in effective way. Despite that the discussion forums give local people the opportunity to talk to each other about the issues that they identify as important, councillors very rarely use blogs and, increasingly, Twitter to share their thinking and update the community on their activities. Councils in Poland very rarely used Facebook fan pages to communicate with the public. Moreover councils all around the country are not getting such results from neighbourhood charters, ward budgeting, planning for real, and other participation techniques such as they can. So when it comes to local dialogue and debate on the Net, the experiences are different in countries with stable democracy and high technological development (e.g., Scandinavia) than in regions of shaping democracy and lower technological impact (e.g., Poland). Analyzing the case of Poland the author sees that communication within a delimited group, for instance a political party organization, gives the best communication results. In contrast, large, open debate forums are more troublesome. If the participants are not committed, the quality of the discussions suffers, and as a consequence the debates die out quickly. Some debates have had a more uncivilized tone than we are used to in public debates, often instigated by anonymous contributions or contributions under false names. Finally, on the negative side, right-wing extremist groups have used some discussion forums to spread racist messages. The Internet's low free speech threshold that should be a clear democratic advantage, can, unfortunately, also be a problem.

Of course, there are also many positive experiences from the Net-based debate. In particular, both scope and quality are affected when citizens have a chance to become involved in specific issues, which the politicians are then forced to deal with. In these cases, the Net functions as an extra channel in addition to meetings, petitions, and so on. The success criterion when it will be assessed in the scope of the Net debate does not have to be 100 percent participation; much less will do. In comparison, only a small minority writes letters to the editor or takes the floor at public meetings. Moreover, some of the participants in the Net debate would never write a letter to the editor or speak out—for the simple reason that they do not subscribe to a newspaper or attend public meetings. Finally, success or failure cannot be "measured" only by the number of active participants, but must also be measured by how many people visit the websites in question and keep informed on the debate (Millard 2006; Porwol, O'Donoghue, and Breslin 2012).

So can ICT further local democracy by creating the new shape of the local public sphere? This question is pertinent in light of the current problems in democratic organization, which are highlighted by two factors. The first one is that the processes of political governance are fragmented and complex and therefore difficult for ordinary citizens to grasp. Second is that the traditional channels, which formed the basis of democratic organization for more than a hundred years, have weakened, and the gap between ordinary people and the political elite seems wider than before. It is therefore natural to search for reforms in democracy and among these to consider the potentials of ICT.

METHODOLOGICAL CONSIDERATIONS

Aims of the Research

When we look closer at the local governance across Europe we can clearly see that different ways of strengthening citizens' trust in local authorities and the creation of platforms of public dialogue are important factors that shape contemporary democracy. Local communities search for solutions that will help in realizating the essential principles of community communication which are public trust and social dialogue. Both of those elements strengthen the sense of responsibility for the region that positively influences the quality of the local governance and all other institutions that create the local public sphere. In the twenty-first century, the century of the digital communication reign, these needs must be realized by creating sufficient digital communication platforms. According to all those aspects the main aim of the paper is to analyze democratic challenges and problems associated with new information and communications technology (ICT) being implemented by a local communication system in Poland. Due to this research the author may have gathered new knowledge about the potentials and limitations of ICT in terms of democratic participation and governance; learned about strategies (adjusted/not adjusted to community expectations) for information society creation and making local politics more intelligible for the citizen; and influenced the communication policy of different regions.

Research Questions

Can ICT further democracy by creating the new shape of the public sphere? This question is pertinent in light of the current problems in democratic organization, which are highlighted by two factors:

1. First, that the processes of political governance are fragmented and complex and therefore difficult for ordinary citizens to grasp;
2. Second, that the traditional channels, which formed the basis of democratic organization for more than a hundred years, have weakened, and the gap between ordinary people and the political elite seems wider than before. It is therefore natural to search for reforms in democracy and among these to consider the potentials of ICT. Those factors bring also more detailed questions for Polish researcher as for example: Whether and how ICT has changed the functioning of local government in Poland? Does ICT support political communication? Is ICT used as platform for local debate? Is ICT used mainly as a tool for transmitting information from the dominant one-sided communication? Does ICT support local democracy?

Technical Aspects of the Research

The paper is based on the qualitative, quantitative research and websites content analysis. For comparison, a survey has been carried out in different regions of Poland. The key aspects for choosing communes were as follows: (1) economy (rural; semiurbanized; urbanized); (2) location (east/west/central part of Poland); (3) population (demography structure). This comparative study was prepared based on (1) documents, reports, statistics data analysis; (2) interviews with politicians, clerks, journalists, entrepreneurs—the purpose of interviews was to determine how increased use of ICT changes work procedures, routines, and involvement in associations internally in the organizations and in relations between citizens and politicians/administration; (3) interviews with inhabitants—representative sample of the citizens in chosen municipalities were asked questions about their knowledge of local politics; participation in associations, institutions, local politics, and so on; perception of influence on local politics; attitudes towards local democracy; use of ICT, and so on. To be able to control for general changes in the citizens' participation and their relationship with the local government, questions were repeated to a random sample of citizens (62) in each municipality; (4) content analysis of chosen municipalities websites was based on parameters that focus on the character of the net communication. Collection and registration of material that focuses on political and community communication—"whole site approach" of both official documents (e.g., administrative documents, official email contact, e-news, e-petitions etc.) and unofficial texts and links (e.g., chat rooms, blogs, social media etc.).

Legitimation of the Methodology

As the digitalization process has diverse courses in different cultures so it is worth doing some deeper studies based on sociological methods. Using both quantitative and qualitative aspects of research the author of this paper is trying to analyze how the power of ICT is used by local communities in regions of different economic level. Is it a big difference between them at this ground? Moreover due to the following methodology we may also check the implementation of ICT tools for building up contemporary democracy in Poland. What is also valuable and possible to check according to methods used by the author of the paper it is the opinion of local society about the impact of ICT over contemporary local communication, electronic administration, and participation processes. Over the past decade local public spheres all over the world were exposed to similar external driving forces such as technological change, internationalization, and concentration of media ownership as well as developments in public administration, so it seems interesting to look at each single case and try to find out how each country is trying to cope with all those challenges.

Research Results: Main Actors of Local Community Life On-line: Municipalities, Local Political Parties, NGOs

The most prevalent form of local communities' online presence are the websites prepared by the municipalities, local political parties, and nongovernmental organizations (NGOs). Research conducted in Polish communes[2] over the last decade proved that they have become more transparent due to their visibility and activity in the network (Guzik 2004; Żygulski 2004; Nowina-Konopka 2009; Biernacka-Ligięza 2010; Nowak 2011). The vast majority of them can be contacted via e-mail, most have their own websites. What is more, the content and level of information provided on the municipal website, is diversified. On the pages of some municipalities, vast amounts of information are provided, while others limit themselves to publishing the address or telephone number. At the beginning many municipalities treated their websites as a hobby or fringe activity, much less important than the ordinary work of the commune. Currently the communication network is becoming one of the most important for many communities and, increasingly, it becomes more professional: more municipalities actively use the Internet, and newly arrived residents are eager to use electronic solutions (Dzierżoniów commune—Customer service assessment).

It is necessary to look closer at smaller communes in different regions of the country—for that analysis there were chosen twelve communes from three different regions of the country: Lower Silesia (western Poland): Wałbrzych, Świdnica, Dzierżoniów, Jaworzyna Śląska; Podkarpacie (eastern Poland): Krosno, Sanok, Jasło, Rymanów; Łódzkie; Voivodship (central Poland): Sieradz, Zduńska Wola, Łask, Zelów (see Table 7.1). We should refer to the presentation of the municipality not only as a political and administrative unit, but also as an institution and organization operating within its area.

All the municipalities covered by the survey have their own websites, most of which are extensive and comprehensive in their provision of detailed information.

This applies to all kinds of information, from driving directions to the school, through cultural events, and ending on the forms to be filled in order to get a place in kindergarten. However, still not many local communities in Poland can apply for a place in kindergarten through the Internet because they have not yet met both the safety and the technical requirements, including the

Table 7.1 Communes Websites Content

	Contact Information	*Council Meetings*	*Online Debates*	*Online Forms*
Wałbrzych	x	x	-	x
Świdnica	x	x	-	x
Dzierżoniów	x	x	-	x
Jaworzyna Śląska	x	x	-	x
Krosno	x	x	-	x
Sanok	x	x	-	x
Jasło	x	x	-	x
Rymanów	x	-	-	x
Sieradz	x	x	-	x
Zduńska Wola	x	x	-	x
Łask	x	-	-	x
Zelów	x	-	-	x

Source: own description 2013.

protection of personal data. Security of the Web-based communication is still a challenge with Polish municipalities, but it has been improving day by day. Municipalities that do not yet offer such a possibility are currently working on it, thus in the near future it is expected that—according to the plan of information society development—sending applications electronically should be possible soon in most Polish municipalities (MSWiA 2013).

Almost all municipalities include on their websites information about the services they offer. The more comprehensive websites (e.g., Sieradz, Wałbrzych) are complex with lots of information. You can find there everything, from the opening hours of the various public institutions such as the swimming pool and library, through information on the operation of kindergartens, the school common room, tax information, the superintendent, planning-town office, or the latest news from the community. Polish website are reminders of cultural events, meetings with artists, and concerts. Residents can also read about the deadlines for applications such as for a place in kindergarten. There is also information about training of the local sports club and the opening hours of the gyms in the municipality.

Information is quite regularly updated (especially in developed communes), and the residents who frequently visit websites of municipalities can keep up with information about public services. They can also easily figure out where to look for more detailed information. Municipal websites also contain links to other sites including institutions and public office sites where you can ask questions on public services.

Many municipalities also publish the results of public opinion surveys carried out by public service sector (e.g., Dzierżoniów commune regularly posts results of each year's customer satisfaction survey).

Websites of the local governments do not engage in party politics, but the content is not fully devoid of political questions because quite often it provides various forms of access to the agendas of their meetings. Convergent for Polish services is the provision of information about what is going on at the meetings of the council, the municipal executive board, and committees, although not in all municipalities is information given in the same degree of detail. There is information about the meetings and agendas of different kinds, and about decisions taken that can also be quickly learnt and accessed from the network. In this respect, small municipalities do not differ at all from the big ones: website users are informed in the same way about the resolutions of the municipality by the small communes of Jasło or Jaworzyna Śląska and the bigger communes as Sieradz or Walbrzych. The difference is, however, in the amount of information. In addition, large communes seem to have more professional websites. Some sites include links to the source of information about provincial council and the local elections.

The list of issues raised at meetings of the council and the municipal executive board and other committees is presented in electronic services, in order to easily find out what issues local government are currently engaged in. But only some municipalities inform in detail about current political issues. However, municipalities are present on the Internet primarily as provider of public services. The political aspect is not as omitted, though it is not readily available in many communes as the information is not posted on the main sites; therefore, to find information about the activities of municipalities, as political organizations, or information about municipal policies users need to either go to the microsite, or go directly to these sites by browser, or commercial information services. The information services of some communes provide a list of politicians, with their phone numbers and e-mail addresses, but you cannot always find the information represented by the political parties. But in most cases politicians are only mentioned as members of the municipal executive board; getting a mention on this list is sometimes difficult. The website of municipalities very rarely have links to local political parties. None of the Polish municipalities analyzed in this paper posted links to the local political parties' websites.

When we look at local parties' websites, we can find information about people to contact and opportunities to establish it with representatives of the parties (here, mostly appears contact to a local party branch chairman.) But we cannot say that this information is open space for dialogue. You can make contact—by phone or via e-mail—with the president or other members of the party. Whether this will be a two-way contact—between a voter and a member of the party or between the citizen and the politician—or whether it contributes to public debate on the Internet, depends on how parties treat such notifications. Unfortunately in Poland, very few local parties creates opportunities for active participation in political dialogue at the local level through the network for the residents. Only in bigger analyzed cities (Wałbrzych, Sieradz, Krosno—former Voivodship center) a local political debate in the network is more visible, but it still does not constitute predominant dialogue space for the local public sphere. It also seems that these forums that exist are not used extensively. Neither do the larger or smaller political parties offer discussion groups or other opportunities that expand the debate on the network. In some municipalities it is the smallest local party or municipal association which offers such function (e.g., Organization for Better Society [OBS]).

In varying degrees, information about offers of different NGOs can be found. Some of these associations have complex websites, others do not have websites at all and are only mentioned in the municipal service. Lists like these are written by the same pattern as the traditional presentation of organizations in the written information issued by local governments such as brochures describing what is happening in the district. This includes both municipality and NGOs' offers.

It follows that all analyzed actors of local public life (municipalities, political parties, NGOs) are able to inform fairly well residents of democratic processes, but they are doing worse with allowing residence to have an active democratic influence. Websites of the main actors of local public sphere have the potential to serve as a quality information source, but they still function as one-way communication from the local governments to the citizens. Attempts to enrich the websites with interactive engines that would enable citizens to use their own voice are rare. When they use the Internet, the most important thing for them is probably not political participation of citizens, but the services that the municipality offers. The actors are focused here on high quality gained by the new technologies—the role of citizens is not so accentuated as "the role of co-owners" (Winswold 2007).

Attempts to increase civic engagement—through discussions or direct decision-making—are rare and do not function very well. An important reason for this is probably the fact that the forums are in no way related to political decision-making processes in communities and that local politicians do not engage in debate.

Where the network is a communication space, it is not municipalities that initiates it but local or regional newspapers with an online version or commercially owned local information services (e.g., www.doba.pl*;* *www.wałbrzyszek.pl*)—sometimes political parties are an exception. Also local portals with discussion forums (e.g., www.walbszyszek.com*;* www.krosno24.pl*;* www.doba.pl*;* www.sieradzanie.pl) can be indicated, where in the local political problems are discussed and Internet users take up important matters. This option is, however, still limited. With regard to the news website, common use of the Internet is something different from using the Internet as a channel of communication between the municipality and residents.

Websites of Local Public Actors as a News Source and an Arena of Participation

At the beginning of this century small municipalities in Poland were subjected to criticism. They were criticized for lack of professionalism, and it was said that they do not even deserve to be called web pages. This criticism even then was not justified. The smallest municipalities did not and still do not differ from the largest in terms of informing the public about services or about the meetings of the municipal council. Differences are primarily focused on the amount of information rather than on their quality, shape, or manner of communication. Large municipalities have more information to share, as such Krosno municipality website is seen to have more content than the Rymanów municipality site. Moreover the shape, form, and quality of the website design is being determined by the commune budget, which unfortunately, in the case of small municipalities is sometimes quite limited. It must be admitted that the quality of web pages in Polish municipalities in 2002 left much to be desired (especially when it comes to small municipalities). The passing decade definitely changed the image of the Polish e-community and the difference between a large and small municipality in the network primarily refers to the quantity rather than quality of the information published on the website of the municipality.

For residents of communities being analyzed in this paper the Internet at the beginning of this century was not a very important source of information about local issues. Only a few people read the mayors' and other politicians' statements that appear in the network (less than 1% of Poles). Only 1% of the population of Poland was actively involved in Internet political debates (Author's own research 2007). Today the situation is slightly different, especially when it comes to the Internet as a source of information, which for many residents of Polish communities is precisely the most important place where they seek information relevant to them about the region (approx. 39%). However, they do not seek such information at the municipal sites but rather use the commercial information services such as www.krosno24.pl.*;* www.swidniczka.pl*;*www.doba.pl*;* www.walbrzyszek.pl and so on. The local newspaper also remains a quite important source of information about the municipality (approx. 31%) (Author's own research 2009). Of course, paper publications are more often accompanied by electronic editions. Today, almost every local title has its own page on the Internet.

Taking into account the opportunity to participate in political discussions on the Internet, it may not be surprising, since the possibilities of discussing politics in the network are few, they are limited and in addition very often linked to specific political parties. Since few residents are members of political parties, we cannot expect that participation in the existing newsgroups will be high.

However, it might have been expected that the number of people who have seen or read posted statements of mayors or other local politicians on the Web, would be much greater. Such statements can be found not only on the municipal website and on the websites of parties, but also, and more often than in the electronic version, in newspapers (only 3% of inhabitants of Poland have become familiarized with these statements through the Internet).

It is worth indicating that in Poland the Internet is a very important source of information about the region, its culture, tourist attractions, economy but most respondents (76% of Poles) (Author's own research 2009) stated that the Internet is still not an important source of information about local politics. Taking into account that over 34% of Poles found that the local newspaper is an important source of information, it can be concluded that the Internet as a source of information about local politics and as an arena to exchange views on political issues, is still in its early stages of development.

If the use of Internet as a source of information in the local context is not very extensive, than the use of Internet as an opportunity for participation in local public life is almost negligible. The possibilities of inviting residents to the political debate in the municipality such as in newsgroups or via the feedback and reactions to the proposals of politicians are not often used in Poland. It seems to be improved, and in the last two years we have found more municipalities where we can meet some positive examples of public dialogue being followed on the Internet (e.g., Dzierżoniów commune is the first municipality in Poland that was granted the administration quality system ISO, and two times a year there is an open dialogue with citizens about service quality at the local administration offices. In the debate with the inhabitants the mayor, high management of the council, politicians, and clerks take part. Participants of the discussion give grades to the administration and suggest some changes which can improve the quality of the local administration service).

CONCLUSIONS

We can see though that local government is leading the way in the use of new technologies and is trying to connect online and offline activity. One thing should be underlined—central government support for good practice in the use of new technologies in local government. Local governments need to use the Web more proactively to engage communities around all strategic issues, but this commitment must be backed by online transparency. Internet proved to be for the majority of analyzed communes mainly a tool for conveying information and providing services on the appropriate level, very rarely used as a platform for debate. It can be noticed that still an important role in building local democracy and local identity continues to be played by local newspaper (in printed and electronic version), where readers can find both information and also have the opportunity to participate in the debate; however, local news sites are becoming even more popular (often through commercial initiative, or additional activity of local publishers/ broadcasters) and usually win with the press.

Unfortunately Polish local politicians are scarcely "visible" on the Internet so it is rather difficult to find information about local political parties which do not provide up-to-date information to their constituency. Poles are rather not looking for information about local politicians/politics on the Internet because they show little interest in this area of local activity; they are more active in the field of sport, education, or culture. So the analysis presented in this chapter brings one very important observation about the lack of Web 2.0 activity in the Polish local politics. Local politicians need to change their communication attitudes, and they should open to social networking technologies to engage citizens in public debate. They need to allow access to social networking sites from council IT systems. It must be borne in mind that there are few hard and fast rules about the use of Web 2.0 technologies. Different tools should be applied in different ways to different problems at different times. For the twenty-first-century politician it is important to be open minded, experimental, and willing to engage in dialogue and to question (or allow others to question) existing hierarchies and ways of working. Achieving successful outcomes will involve uncertainty and productive chaos along the way, and it is essential to adopt a mindset that is comfortable with this way of working with people. Collaboration is becoming a more vital currency than control. In time it will become redundant to talk about "using the Web" as these techniques and tools become fully integrated into the way in which councils work. But it's vital to mainstream the mindset as well as the technology and ensure that the best aspects of the Web—open, collaborative, non-hierarchical and thus, ultimately, democratic—are central to all our activities online and off.

To sum up, one has to conclude that the Internet can contribute to the reconstruction of the local public sphere because it is a convenient way of communication. Any content can be put online, and it does not require special skills. In a situation in which the local media lose their meaning, the Internet could fill the gap. It also gives an opportunity to participate in a bigger debate into which more people are engaged than would be the case with the TV or newspapers. The Internet also creates a possibility to have a debate on socio-political issues and to organize political marketing online. However, it is rarely used, especially in small towns and villages. Despite social changes, technology development, and chances of participation, the analysis shows that local communities rather slowly adjust to these changes, they do not organize initiatives, and they do not fully realize the potential of the Internet.

NOTES

1. Web 2.0 is a term coined by O'Reilly Media in 2004. The term Web 2.0 has many definitions, and includes the use of tools such as social networking sites, videosharing sites, blogs, and wikis. However, what makes Web 2.0 new and important is its capacity to change the relationship between the Internet and its users. According to Lilleker and Jackson, Web 2.0 is about interacting with Web-based content, adding comments, or uploading files. Each visitor is able to have shared ownership over a site. This implies a change in power structures and a shift in organizational thinking toward models based on equal partnership rather than elite dominance (Lilleker and Jackson 2010).

2. "commune" and "municipality" are both terms used to name Polish "gmina" (one of the level of the Polish local government system). The term "commune" has

more geographic and organization connotations and "municipality" is rather interpreted as a more adminstrative term more often like 'commune administration sector." "Municipality" is dedicated to city/town commune.

REFERENCES

Anderson, B. 2006. *Imagined Communities: Reflections on the Origin and Spread of Nationalism.* London: Verso.

Appadurai, A. 1996. *Modernity Al Large: Cultural Dimensions of Globalization.* Minneapolis: University of Minnesota Press.

Barney, D. 2004. *The Network Society. Key Concepts Series.* Cambridge: Polity Press.

Bell, C. and H. Newby. 1976. "Communion, Communialism, Class and Community action: The sources of new urban politics." In *Social Areas in Cities Vol. 2*, ed. D. Herbert, R. Johnston. Chichester: Wiley.

Białobłocki, T., J. Moroz J., and M. Nowina-Konopka. 2006. *Społeczeństwo informacyjne: istota, rozwój wyzwania.* Warszawa: Wydawnictwo Akademickie i Profesjonalne

Biernacka-Ligięza I. 2010. "Local communication matters?—local information services in the creation of the civic society—the case of Poland." Paper presented at the *IAMCR Conference: Communication and Citizenship: Rethinking Crisis and Change*, Braga, July 2010.

Chadwick A. 2006. *Internet Politics: States, Citizens and New Communication Technologies.* New York: Oxford University Press.

Cindio F. and D. Schuler. 2012. "Beyond Community Networks: From local to global, from participation to deliberation." *The Journal of Community Informatics* 8(3). http://www.ci-journal.net/index.php/ciej/article/view/908/933.

Coe, A., G. Paquet and J. Roy. 2001. "E-Governance and Smart Communities—A Social Learning Challenge." *Social Science Computer Review* 19(1):80–93.

Cooper, P. and P. Kulawczuk, eds. 2008. *Electronic Self-Government Communication. Innovative Tools of Public Interest Advocacy.* Warszawa: Wydawnictwo Akademickie i Profesjonalne.

Cullen, R. and L. Sommer. 2010. "Participatory Democracy and the Value of Online Community Networks: an exploration of online and offline communities engaged in civil society and political activity." *43rd Hawaii International Conference on System Sciences.*

Dahlgren, P. 2009. *Media and Political Engagement: Citizens, Communication, and Democracy.* New York: Cambridge University Press.

Featherstone, M., S. Lash and R. Robertson. 1995. *Global Modernities.* London: Sage.

Fountain, Jane E. 2001. *Building the virtual state: information technology and institutional change.* Washington: Brookings Institution Press.

Giritli, Nygren K. 2009. "E-Governmentality: on Electronic Administration in Local Government." *Electronic Journal of e- Government* 7(1), 55–64.

Grönlund, Å. and A. Andersson. 2007. "E-government Research 2003–2006: Improvements and issues." In *Current Issues and Trends in E-Government Research,* ed. D. Norris, 247–288. Hershey: Idea Group.

Guzik, R. 2004. Polskie miasta i gminy wiejskie w Internecie—geografia społeczeństwa informacyjnego. In. *Rozwój miast i zarządzanie gospodarka miejską. Miasta w okresie przemian*, ed. J. Słodczyk. Opole: Wydawnictwo UO.

Habermas, J. 1989. *The Structural Transformation of the Public Sphere: An Inquiry into a Category of Bourgeois Society*. Cambridge: Polity.

Hafez, K. 2007. *The Myth of Media Globalization*. Cambridge: Polity Press.

Hargittai, E. and G. Walejko. 2008. "The Participation Divide: Content Creation and Sharing in the Digital Age." *Information, Communication and Society* 11(2), 239–256.

Haug, A. V. 2007. "Local Democracy Online: Driven by Crisis, Legitimacy, Resources, or Communication Gaps?" *Journal of Information Technology And Politics* 4(2), 79–99.

Hegde, R. S. 2005. "Disciplinary Spaces and Globalization: A Postcolonial Unsettling." *Global Media and Communication* 1(1), 59–62.

Held, D. 2006. *Models of democracy*. Cambridge: Polity Press.

Hoff, J., I. Horrocks and P. Tops, eds. 2000. *Democratic Governance and New Technology. Technologically Mediated Innovations in Political Practices*. London and New York: Routledge.

Ifinedo, P. and M. Singh. 2011. "Determinants of eGovernment Maturity in the Transition Economies of Central and Eastern Europe." *Electronic Journal of e-Government* 9(2), 166–182.

Lilleker, D. and N. Jackson. 2010. "Towards a More Participatory Style of Election Campaigning: The Impact of Web 2.0 on the UK 2010 General Election." *Policy & Internet* (2), 69–98.

Jeitziner, B. 2009. "Zielsetzung und Umsetzung eines webbasierten Wahlhilfesystems für Parlamentswahlen." In. *E-democracy & eGovernment, Stages of a Democratic Knowledge Society,* ed. A. Meier, 178–183. Berlin: Springer.

Kies, R., F. Mendez, P. Schmitter and A. Trechsel. 2004. "Evaluation of the Use of New Technologies in Order to Facilitate Democracy in Europe." *STOA 116. European Parliament Report*. www.europarl.europa.eu/stoa/publications/studies/stoa116_en.pdf

Klotz, R. J. 2004. *The Politics of Internet Communication*. Lanham: Rowman & Littlefield.

Kluszczyński, R. 2001. *Społeczeństwo informacyjne. Cyberkultura. Sztuka multimediów*. Kraków: Rabid.

Layne, K. and J. W. Lee. 2001. "Developing Fully Functional E-Government: A Four Stage Model." *Government Information Quarterly* 18(2), 122–136.

Lee, J. and S. Kim. 2012. "Active E-participation in Local Governance: Citizen Participation Values and Social Networks." *EGPA Conference*.

Luke, Timothy W. 1995. "New World Order or Neo-World Orders: Power, Politics, and Ideology in Informationalizing Glocalities." In *Global Modernities*, ed. M. Featherstone, S. Lash and R. Roberston, 91–107. London:Sage.

Macintosh, A. and A. Whyte. 2008. "Towards an Evaluation Framework for eParticipation." *Transforming Government: People. Process & Policy* 2(1), 16–30.

Meier, A., ed. 2009. *E-Democracy & E-Government: Stages of a Democratic Knowledge Society*. Berlin: Springer.

Millard, J. 2006. *E-Governance and E-Participation: Lessons from Europe in promoting inclusion and empowerment.* Budapest: UDESA.

Morrison, J. and D. Newman. 2001. "Consultation and Participation in New Labour's Britain and Beyond." *International Review of Law* 15(2),171–194.

Nam, T. 2010. "Whither Digital Equality? An Empirical Study of the Democratic Divide." *43rd Hawaii International Conference on System Sciences.*

Norris, P. 2001. *Digital Divide: Civic engagement, Information Poverty and the Internet Worldwide.* New York: Cambridge University Press.

Nowak, J. 2011. *Aktywność obywateli online*. Lublin: UMCS.

Nowina-Konopka, M. 2009. Politycy lokalni a nowe typy komunikacji politycznej w Internecie. In *Nowe media i polityka*, ed. M. Jeziński. Toruń: Wydawnictwo Adam Marszałek.

Orihuela, L. and T. Obi. 2007. *E-government and E-governance: Towards a Clarification in the Usage of both Concepts. E-governance: A global perspective on a new paradigm.* Amsterdam: IOS Press.

Papadopoulou, P., M. Nikolaidou and D. Martakos. 2010. "What is Trust in E-Government? A Proposed Typology." *43rd Hawaii International Conference on System Sciences.*

Porwol, L., P. O'Donoghue and J. G. Breslin. 2012. "Social Inclusion and Digital Divide: eParticipation Dilemmas in Municipalities." *The 6th International Conference on Theory and Practice of Electronic Governance.*

Pratchett, L., M. Wingfield and R. K. Polat. 2006. "Local Democracy Online: An Analysis of Local Government Web Sites in England and Wales." *International Journal of Electronic Government Research* 2(3), 75–92.

Putnam, Robert D. 2001. *Bowling Alone: The Collapse and Revival of American Community.* New York: Simon & Schuster.

Robertson, R. 1995. Glocalization: "Time-Space and Homogeneity-Heterogeneity." In *Global Modernities*, ed. M. Featherstone, S. Lash, and R. Roberston, 25–44. London: Sage.

Romsdahl, R. J. 2005. "Political Deliberation and E-Participation in Policy-Making." *Comparative Literature and Culture* 7(2), 2–11.

Rose, L. E. 2002. "Municipal Size and Local Non Electoral Participation: Findings from Denmark, the Netherlands, and Norway." *Government and Policy* 20, 829–851.

Thornton, W. 2000. "Mapping the ''Glocal'' Village: The Political Limits of 'Glocalization'." *Continuum: Journal of Media & Cultural Studies* 14, 79–89.

Thussu, D. K. 1998. *Electronic Empires—Global Media and Local Resistance*. Arnold: London and Oxford University Press, New York.

Torpe, L. and J. Nielsen. 2004. "Digital Communication between Local Authorities and Citizen in Denmark." *Local Government Studies* 30(2), 230–244.

Trechsel, A. H. and F. Mendez. 2005. *The European Union and E-Voting*. London: Routledge.

Urry, J. 2000. *Sociology Beyond Societies. Motilities for the Twenty-First Century.* London:Routledge.
Wilhelm, A. 2000. *Democracy in the Digital Age.* London: Routledge.
Wilhelm, A. 2004. *Digital Nation: Toward an Inclusive Information Society.* London: Cambridge University Press.
Winsvold, M. 2007. "Municipal Websites in the Local Public Debate: Supplying facts or setting agenda?" *Nordicom Review* 2 (2007), 7–23.
Yates, D. J., G. J. Gulati and A. Tawileh. 2010. "Explaining the Global Digital Divide: The Impact of Public Policy Initiatives on Digital Opportunity and ICT Development." *43rd Hawaii International Conference on System Sciences.*
Zissis, D. and A. Papadopoulou. 2009. "Competent Electronic Participation Channels in Electronic Democracy." *Electronic Journal of E-government* 7(2), 195–208.
Żygulski, W. 2004. E-govrnment-rozwiązanie przyszłości. *Marketing* 10(3), 2–12.

SOURCES

Author's survey: followed between 2002–2010: place of survey: former Walbrzyskie Voivodship (counties: Dzierżoniowski; Świdnicki; Wałbrzyski); quantitative questionnaires between inhabitants 2000 (based on telephone survey); qualitative interviews with local politicians; officials; local journalists; representatives of NGO organizations—number (34);
*Reports:*Administracja publiczna w sieci 2002—czy rzeczywiście bliżej obywatela (2002) Internet Obywatelski, Warszawa.
eEuropa—Społeczeństwo informacyjne dla wszystkich (2000) Monitor Integracji Europejskiej nr 35, Urzędu Komitetu Integracji Europejskiej, Warszawa.
Strategia rozwoju społeczeństwa informacyjnego, (2008) Ministerstwo Spraw Wewnętrznych i Administracji, Warszawa.
Rozwój regionalny w Polsce—raport (2009). Ministerstwo Rozwoju Regionalnego, Warszawa.
United Nations. 2003. "World Public Sector Report 2003: E-government at the Crossroads." ST/ESA/PAD/SER.E/49, New York.
MSWiA. 2013. "The Strategy for the Development of the Information Society in Poland until 2013."
Statistics: 2002–2011 reports of: Central Statistics Office (GUS); Eurostat; Internet world statistics; Millward Brown SMG/KRC; PBI; Gemius; municipalities presented in the paper.

Chapter 8

Youth Vote and Web 2.0 Political Engagement

Can Facebook Increase the Propensity of Young Citizens to Vote?

Alan Steinberg

According to a recent study, 85% of college students have engaged in some form of online political participation. Yet when asked about more traditional forms of political participation, that statistic drops to 59%.[1] This implies that over 25% of young people surveyed are participating politically, but through a means that traditional metrics and most surveys on political behavior fail to capture. This is a significant difference! If we as political scientists are not even aware of this happening, how can we start to understand this effect on our political system?

The 2008 elections might have provided the first mainstream use of social media in federal campaigns, but individuals have likely been making use of online social media for political discourse and participation since even before that time. Since the advent of Facebook, and even before with blogs, chat-rooms, and other means of online participation, an increasing number of people are accessing the Internet not only to obtain political knowledge, but to interact as part of the Web 2.0 political community. Recent work has helped clarify who is using the Internet for "cyber participation," and its effect on turnout (Steinberg 2010).

This study seeks to advance upon previous work to determine how the determinants of cyber participation compare to those of other forms of online participation as well as traditional participation. First, this study will show that cyber participation is not based on traditional socioeconomic factors, and unlike traditional participation is not driven by political drive or ideology, but rather by political curiosity and engagement. Second, this study will demonstrate that even when controlling for other forms of participation, political engagement though Web 2.0 technologies will lead to increased intentions for youth voting. When these two aspects are taken in tandem, it demonstrates the potential for political science courses to encourage political participation and effect voter turnout.

POLITICAL PARTICIPATION ONLINE

Over the last decade, there has been an increasing amount of work done to understand the effect of the Internet on political participation (Bimer 1998; Tolbert and McNeal 2003; Gibson, Lusoli, and Ward 2005; Best and Krueger 2005; Steinberg 2010). As computers become cheaper, more available, and part of our daily lives access to the Internet only increases. Potential young voters are surrounded by the ability to access the Internet, given home computers, smartphones, computers at school, workplace computers, and public access systems in libraries. Additionally, as the study discussed here was conducted through the Internet, we can assume that college students have nearly ubiquitous access to the Internet today!

While an increasing number of people are using the Internet as a source of news and information about politics, it is not clear to what extent the Internet effects political knowledge (Gronlund 2007). In this study we explore political knowledge, but are not able to address causality. So the question remains, is political knowledge driving political participation online or is online political engagement increasing political knowledge?

Recent research in the field of online participation seeks to identify the difference between various means by which a person can engage politically using the Internet. Steinberg identifies a new form of participation, "cyber participation," which is characterized by "involvement of actors within the Web 2.0 space." He argues that "A Web 2.0 site allows users to interact with other users, in contrast to non-interactive sites where users are limited to passive viewing of information" and that "[i]n this world, web applications are designed to facilitate interactive information sharing and collaboration" (Steinberg 2010).

We know that users of these sites are writing comments, joining interest groups, and discussing politics in ways that were previously not measureable. This cyber political discourse may also lead to other forms of civic participation (Klofstad 2007; Kobayashi, Ikeda, and Miyata 2006; Steinberg 2010). However, by some metrics, the age group that is the most active within the Web 2.0 community is also considered to be more politically disengaged than older generations (Zukin et al. 2006). It is for this reason that it is so crucial to examine the effects of cyber participation on the potential youth vote.

In political science, we generally accept that the SES model, based on components of socioeconomic status, such as education and income, does a good job at predicting political participation (Brady, Verba, and Schlozman 1995). But previous research suggests that the SES model fails to predict cyber participation (Steinberg 2010). This study shall examine if either of these findings holds true for young citizens especially when differentiating traditional, online, and cyber participation.

HOW TO MEASURE CYBER AND ONLINE POLITICAL PARTICIPATION

Defining online political participation is relatively straight forward. In short, online participation is any form of political participation that can take place online. For example someone can donate money online, sign up to receive updates about campaigns online, research information about a candidate online, or watch a video about the campaign online. While these activities are happening on the Internet, there are clear parallels to traditional participation. Cyber participation on the other hand differs as it is a new outlet rather than a modernized version of traditional participation.

Cyber participation is not the same as traditional participation, nor is it the same as traditional "online participation." Studies are quick to link aspects of Internet participation to similar modes of traditional participation. This includes reading campaign literature online, contacting an elective representative online, contributing to a campaign online, or signing a petition online (Best and Krueger 2005; Smith et al. 2009). Based on using these metrics it is no surprise that findings suggest that online participation is similar to that of traditional offline participation.[2]

Cyber participation is different from online participation. Social networking sites provide a platform for people to discuss ideas deliberate politics, and perhaps even for persuasion to take place. Within these networks there is the ability to create political communities that can provide a fluid and dynamic response to issues, quickly and efficiently (Bimer 1998). While there is no disagreement that it is important to study online participation, with an increasing number of people joining social networking websites, cyber participation will quickly become an important metric. Within social networking sites, individuals can become "*friends*" with a politician or candidate, initiate or join a political themed group, or post questions or comments about an election.

These actions are about engagement within a community. They allow an individual participant to have a broad reach within their social network and influence multiple aspects of political participation (Steinberg 2010). Such reach is important as generally people only maintain active ties to a subset of their acquaintances. Social networking websites help people keep their ties to others more active as they facilitate forming new connections between people and communities (Wellman et al. 1996).

Both online and cyber participation requires very little in the way of resources, both being less time-consuming and having lower costs of engagement. Despite this, both of these means of engagement can yield positive effects in the realm of political participation.

WHY YOUNG PEOPLE ARE MORE LIKELY TO BE ONLINE OR CYBER PARTICIPATORS

Smith et al (2009) find that participation, both online and off, is still in the domain of those with higher income and higher education. However, the Internet is becoming increasingly used as a forum for political debate in both the United States and abroad with the potential for fast and unhindered communication among potential voters (Jensen 2003). This being said, even whether or not online political participators differ categorically from traditional participators is still up for debate.

Other research has shown that the factors predicting online and cyber participation differ from those of offline participation (Best and Krueger 2005; Steinberg 2010). While Best and Krueger (2005) identify some differences, they still find that the same SES determinants are present in online participants, suggesting that at some fundamental level perhaps the users are the same. Steinberg (2010) on the other hand argues that due to the characteristics of cyber participation itself, those who participate though online social media are different from traditional participators as they represent a broader range of the public at large.

For young voters we would expect more online and cyber participation due to the lower economic and time costs as compared to more traditional forms of participation. For example, younger voters are less likely to have disposable income to donate to political parties or candidates. Additionally, they may lack the time to attend a political rally but rather attend the same style rally in an online environment such as Second Life, or some other form of online town-hall style event that is more convenient.

Online and cyber participation both also mitigate other socioeconomic characteristics that may curse more traditional forms of political participation. For example, gender and race could hardly serve as a barrier as the Internet is open to anyone. Meanwhile, online social networking makes finding like-minded people to network with politically easier than ever before. In addition, cyber participation also is generally localized to a person's existing online network, and thus their participation engages people already tolerant of each other, and increased interaction leads to increased tolerance (DiMaggio et al. 2001; Kobayashi, Ikeda, and Miyata 2006).

Political knowledge is also likely to play less of a role for online or cyber participation as the threshold for engagement is low. A person does not need much political knowledge to start asking questions or posting comments. There are a multitude of opportunities, and the interactive aspects will lead to a quick growth in political knowledge should someone wish to take advantage of it (Gronlund 2007). Such interactions may even lead to more traditional offline participation as individual political engagement grows (Klofstad 2007).

DO YOUNG ONLINE AND CYBER PARTICIPANTS TURN OUT?

The traditional "so what?" question pops up when talking about young participants; regardless of their engagement, does this engagement actually translate to voting? If these new forms of participation are eliciting a different group or even just engaging potential young voters in ways not seen in the past, then it is important to know if this group will turn out.

In 2000, about 74% of 18–29-year-olds were using the internet, but by 2009, 93% of 18–29-year-olds were online, and today almost everyone in this age group uses the Internet. The internet allows candidates to reach the masses easily and cheaply, and with regard to potential young voters, reach them in the world they constantly engage in, and where they are spending their time. In addition, through the Internet, both new and more traditional forms of participation have become easier and in theory accessible to a greater number of people.

The impact of Internet access and online participation is not clear. Tolbert and McNeal (2003) find that respondents with access to the Internet and online election news were significantly more likely to report voting in the 1996 and 2000 presidential elections, controlling for socioeconomic factors. However, these findings may be attributable to political knowledge given that Internet access itself allows for a person to become more informed about issues and elections and in a quick and efficient manner. Instead of measuring access, more recent research has examined the effects of online participation (Gibson, Lusoli, and Ward 2005; Williams and Gulati 2008; Williams and Gulati 2009). However, this research only looks at part of the story. Gibson et al. show similarly to Best and Kruger that there are some differences between online and offline participators. Meanwhile, Williams and Gulati show that there is an association between social media-based participation and election outcomes. Neither however takes the next step, as will be done here, to link

Internet-based participation to voting, while controlling for other factors that are known to influence turnout.

HYPOTHESIS

This study delves into three general questions. First, are young people more likely to engage in online or cyber participation as opposed to traditional participation? Second, are the predictors of these forms of participation the same? Third, what is the effect of the various forms of participation on a young citizens' intention to vote?

This study seeks to demonstrate that young people are active political participants, but that due to participation online and through Web 2.0 technologies, more traditional modes of surveying political participation, such as those found in the American National Election Study, are unable to detect this participation. Additionally, these online and Web 2.0 forms of participation stem from different underlying factors than traditional participation, due to their lower costs and ease of access, making them an equalizing force for young voters against the bias of traditional participation toward older and better socioeconomically endowed individuals. In turn, cyber participation will lead to a positive effect on turnout due to its ability to engage young citizens in ways similar to how traditional participation leads to turnout.

The following are specific hypotheses that this study will test:

- Hypothesis 1: Young people are more likely to engage in online or cyber participation than traditional participation.
- Hypothesis 2: Socioeconomic factors that generally predict traditional participation will not influence online and cyber participation.
- Hypothesis 3: Online and cyber participation will be significantly associated with the intention of young citizens to vote, even controlling for traditional participation.

Unlike traditional participation, which is resource heavy, the low amount of resources required for cyber participation will allow for unbiased, that is, increased youth, participation. Therefore it is expected that socioeconomic factors such as income, and predisposing factors such as gender and race, will not be significant predictors of cyber participation among young citizens. Additionally, prior research suggests access to the Internet alone increases the probability of voting (Tolbert and McNeal 2003). With this in mind, cyber participation should be related to turnout in ways similar to traditional participation.

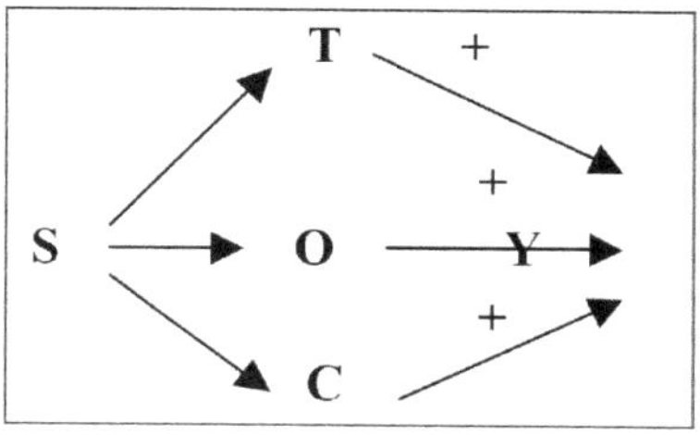

S = Socioeconomic Factors & Party Intensity
T = Traditional Participation
O = Online Participation
C = Cyber participation
Y = Turnout

Figure 8.1

MODEL

Figure 8.1 presents the model being tested in this study. The model suggests that socioeconomic factors and party intensity (S) influence traditional participation (T), online participation (O), and cyber participation (C). Additionally, traditional participation (T) and online participation (O) may influence cyber participation (C). The statistical models presented will show that even when controlling for traditional participation and online participation, cyber participation will still have a positive significant influence on intentions to vote in an upcoming election.

DATA

Given that most national data sets lack information on traditional, online, and cyber participation metrics on the same sample, a survey was designed and conducted in the fall of 2011 on students taking introductory political science courses at the University of Houston for this purpose. This survey combined elements of the Pew Internet and American Life Survey regarding online and social media-based activities, with elements form the American National Election Survey (ANES) regarding political participation. Taken together, this survey allows for measuring each of the three potential modes of participation among the same group of subjects. Students were offered extra credit in their course should they choose to take the online survey. Out of the approximately 2,900 eligible students, 1,605 successfully completed the survey. The students, while of similar age and educational standing, are politically, racially, and economically diverse.

In this analysis political participation is measured in four different ways: cyber participation, online only participation, online participation, and traditional participation. Cyber participation is operationalized in the same way it is in Steinberg (2010), where the respondent is coded 1 or 0 based on if they engage in any of the three forms of cyber participation on an online social networking website: (1) *friending* or *liking* a political candidate, (2) starting or joining a political group, or (3) posting a question or comment about politics.

Online only participation is evaluated dichotomously based on if a respondent engaged in any of the following six activities that could only take place in an online environment and for which there is a limited ability to do similar actions offline: (1) sharing photos, videos, or audio files online that relate to the campaign or the elections, (2) forwarding someone else's political commentary or writing to others, (3) forwarding someone else's political audio or video recordings to others, (4) setting up news alerts to get political or campaign information emailed to you when new information is cited in the news or on the Web, (5) customizing a Web page to display new political or campaign information that is especially interesting or important to you, (6) subscribing to receive campaign or political information through an RSS feed.[3]

Online participation more generally includes six more activities in addition to online only participation. These additional activities are more reminiscent of traditional participation and/or online information gathering rather than active engagement, including: (1) signing up online to receive updates about the campaign or the elections, (2) contributing money online to a candidate running for public office, (3) looking for more information online about candidates' positions on the issues or voting records, (4) watching a video online made by a candidate, campaign, or news organization, (5) watching a campaign- or election- related video online that did not come from a campaign or a news organization (i.e., fan-made or satire, YouTube or Vimeo video), (6) signing up online for any volunteer activities related to the campaign—like helping to register voters or get people to the polls.

Traditional participation is evaluated dichotomously based on engagement in any form of participation as measured by the American National Election Survey, and including (1) talking to any people and trying to show them why they should vote for or against one of the parties or candidates, (2) attending any political meetings, rallies, speeches, dinners, or things like that in support of a particular candidate, (3) wearing a campaign button, (4) putting a campaign sticker on your car, (5) placing a sign in your window or in front of your house, (6) working or volunteering for a political party or candidate, and (7) giving money to a political candidate or party.

In order to understand the characteristics associated with the various means of political participation, this study takes into account socioeconomic factors that have been known to influence traditional and online participation, as well as turnout, as control variables. The following variables about the respondents are considered: gender, race, income, socialization, political knowledge, party identification, and party intensity. Their use will allow for comparisons with traditional political participants as well as general measurements of online participation. Further explanation for the rationale behind controlling for individual variables is explained later in the chapter. While there is no need to control for party identification, it is included in the model in an attempt to gain some insight into which party is making greater inroads in regards to the utilization of new technology via cyber participation. Despite that this question is not directly related to the study, a significant finding might encourage further academic research into this important question.

There is little reason to suspect that gender will play a role in either online or cyber participation, and subsequent turnout; however, due to its inclusion in other models for participation and turnout used in previous research, gender is controlled for here as well. Previous research suggests that political participation and turnout may differ between the sexes, especially when comparing online and offline participation, though no rationales are offered for the differences (Best and Krueger 2005; Tolbert and McNeal 2003; Leighley and Nagler 1992).

In order to see the effect of minority status on cyber participation and turnout, white respondents are coded 1 and non-white respondents are coded 0. While race has been a significant predictor for turnout, it has a limited association with online participation where some research shows it to have a significant effect, but other studies do not (Leighley and Nagler 1992; Tolbert and McNeal 2003; Smith et al. 2009). Tolbert and McNeal show that race itself is not associated with online participation, while Leighley and Nagler show a strong negative association between minorities and turnout. As more minorities gain Internet access and reap the benefits that Internet access may provide, perhaps the effect on turnout will be mitigated. It is not expected that race would be a predictor of youth online or cyber participation in this study as it was not a predictor of either in previous research (Tolbert and McNeal 2003; Steinberg 2010).

Household income is often a predictor of political engagement and mobilization as people with more income can afford to contribute to political causes as well as afford the time to engage in other forms of participation. Income has previously been shown to have an association with online participation (Tolbert and McNeal 2003; Smith et al. 2009). Given the ease of access to computers and the Internet in general, there is no reason to believe that income would be directly related to online social networking usage. Additionally, since the subjects are students, household income may not directly affect their individual spending abilities. Household income is coded as follows: 0 = <$10,000, 1 = $10,000<$20,000, 2 = $20,000<$30,000, 3 = $30,000<$40,000, 4 = $40,000<$50,000, 5 = $50,000<$75,000, 6 = $75,000<$100,000, 7 = $100,000<$150,000, and 8 = $150,000+.[4]

Previous research demonstrated that declining civic engagement could be the cause for reduced turnout (Putnam 2000). Therefore, some sort of control for civic engagement is important. For this study, a measure of church attendance is used to estimate a general civic engagement. While not the best metric for this, religiosity scores have been shown to be generally associated with other forms of civic engagement, and church attendance itself has been show to be a significant factor in predicting turnout (Verba, Schlozman, and Brady, Voice and Equality: Civic Voluntarism in American Politics 1996; Timpone 1998). It has been coded from 0 to 5, increasing as church attendance increases.

Party identification is coded 0 to 5 for Democrat, Democratic leaning, Independent, Republican leaning, and Republican, respectively. This is included to control for any possible party effects as well as to see the degree party matters in regards to the youth intention to vote. Partisan intensity, meanwhile, created by folding the party identification score, coded 0 for independents, 1 for weak party identification, and 2 for strong association is much more important theoretically. Prior research suggests that strong party affiliates are more likely to engage in political participation (Verba and Nie 1972; Best and Krueger 2005). Controlling for partisan intensity allows for parsing out if these new forms of participation are just a new medium for traditional participators or actually engaging a new demographic.

Political knowledge is also included as a control variable. As age and education are typically associated with political knowledge, and given that those two factors are relatively constant for those surveyed, political knowledge serves as a similar substitute. It is expected that those with higher degrees of political knowledge would be both more likely to participate as well as more likely to vote. However, as mentioned previously, it is difficult to tease out the causality of if political knowledge leads to online and cyber engagement or if political knowledge is gained from the engagement itself. Political knowledge is measured on a scale of 0–10 based on the number of correct answers to a series of questions including identification of foreign and local political leaders, state and federal policy issues, government spending, and a question about Facebook. On average respondents answered 7 out of 10 questions correctly, implying a relatively high degree of knowledge.

Two more control variables are added to the analysis of turnout a measurement of trust in government, and overall satisfaction with "the way things are." These metrics have been shown to influence voting in the past, but should not influence political participation in any particular way. Those with less trust in government may not bother to vote, while respondents who are less satisfied may be more likely to want to vote in an attempt to changing the status quo. Both variables are measured on a scale of 0 to 3, from lowest to highest degree of trust and lowest to highest degree of satisfaction, respectively.

METHODS

In order to explore these hypotheses, two distinct, but interconnected, sets of statistical models are used. The first set of probit models examines the characteristics of respondents to determine the factors that influence various forms of participation: traditional, online, and cyber. The second set of probit models examines the association between these various modes of participation and turnout.

The first model is a measure of general cyber participation, coded 1 if the subject engaged in any of the three forms of cyber participation. The second and third models accesses the predictors of online only and general online participation; coded 1 if the subject engaged in any aspect of online participation that does not have a traditional counterpart or any form on online participation, respectively. The fourth model is a measure of traditional participation, coded 1 if the subject engaged in any of the standard traditional participation methods surveyed by the ANES and similar studies. The models are as follows:

1. Cyber Participation = B(gender) + B(race) + B(income) + B(socialization) + B(party identification) + B(partisan intensity)
2. Online Only Participation = B(gender) + B(race) + B(income) + B(political knowledge) + B(socialization) + B(party identification) + B(partisan intensity)
3. Online Participation = B(gender) + B(race) + B(income) + B(political knowledge) + B(socialization) + B(party identification) + B(partisan intensity)
4. Traditional Participation = B(gender) + B(race) + B(income) + B(political knowledge) + B(socialization) + B(party identification) + B(partisan intensity)

Then a series of probit models are used to examine the treatment effects of online and cyber participation on turnout. The models are as follows:

5. 2008 Turnout = B(cyber participation) + B(online participation) + B(traditional participation) + B(gender) + B(race) + B(income) + B(political knowledge) + B(socialization) + B(party identification) + B(partisan intensity) + B(trust in government) + B(satisfaction)
6. 2010 Turnout = B(cyber participation) + B(online participation) + B(traditional participation) + B(gender) + B(race) + B(income) + B(political knowledge) + B(socialization) + B(party identification) + B(partisan intensity) + B(trust in government) + B(satisfaction)
7. 2012 Intention to TurnOut = B(cyber participation) + B(online participation) + B(traditional participation) + B(gender) + B(race) + B(income) + B(political knowledge) + B(socialization) + B(party identification) + B(partisan intensity) + B(trust in government) + B(satisfaction)

RESULTS AND ANALYSIS OF CAUSES OF PARTICIPATION

Out of the 1,605 respondents, 945 survey respondents (58.88%) engaged in some form of traditional participation, 904 (56.32%) engaged in one of the three types of cyber participation, and 1,366 (83.24%) engaged in some form of online participation. While the difference between traditional and cyber participation is only 2.55%, and thus not statistically significant, the difference between traditional and online only participation is 24.36% which is a statistically significant difference at the $p < .05$ level. This demonstrates that the future generations of potential voters are more likely to engage through an Internet-based medium than though traditional participation.

The most common form of traditional participation was "Talked to any people and try to show them why they should vote for or against one of the parties or candidates," a form of traditional participation that could easily take place through online social networking. 30% of respondents engaged in this form of traditional participation. The way the survey was written, there is no way to be sure that this communication happened in a traditional setting, such as in person or on the phone, and not via online social networking such as through Facebook.

In order to see if the respondents engaging in each form of participation are the same a series of correlations was calculated. The correlation between online and traditional participants is .3405, the correlation between online and cyber participants is .3515, and the correlation between cyber and traditional participants is .2929. Additionally, 493 respondents (30.72% of the sample) engaged in either online or cyber participation, but not in traditional forms of participation. So while there is a clear relationship between forms of participation, the various forms also elicit specific users.

The probit results from the first four models examining factors of various forms of participation can be found in Table 8.1. In line with the second hypothesis, traditional socioeconomic factors do not predict online or cyber participation. Gender, race, and income all failed to consistently reach statistical significance at the $p < .05$ level. This implies that the role of these factors in determining online and cyber participation is quite low, and that highlights that these forms of participation do not require the resources as other forms of participation. However, these socioeconomic variables are also not significant in predicting traditional participation. This implies that the socioeconomic effects are a thing of the past as previous research shows their general effect on the population, but this study does not show the effect for this younger generation.

Across the board, income fails to reach statistical significance. Race reaches the $p < .1$ level of significance, but the coefficient is in the wrong direction, suggesting non-white respondents are more likely to engage in online only and traditional participation. Gender is only significant in the online only model, suggesting men are more likely to engage in this type of participation.

The only consistent relationships worth noting are religiosity, party intensity, and political knowledge. Respondents who attend a place of worship more often are also more likely to engage in all three forms of participation, though the effect is most significant for traditional participation. Respondents with high levels of partisanship, are more likely to engage in participation across the board, but only significant at the $p < .05$ level for online and traditional participation. Being a strong partisan identifier implies a young adult is more likely to engage politically, however, the effect is not a strong predictor in regards to cyber and online only participation, but rather for more traditional means of participation. In other words, cyber and online participators (who avoid traditional participation metrics) are less partisan than their traditional participation counterparts.

Political knowledge has an interesting relationship with participation because there is no means by which to evaluate causality. What the data show is that there is a significant association between cyber participation and political knowledge as well as online participation and political knowledge, but that relationship is not significant between traditional participation and political knowledge. Perhaps those who engage through cyber and online participation are gaining their knowledge through engagement, while traditional participants are not building knowledge through engagement. Another possibility is that more politically knowledgeable individuals are also just more likely to engage through cheaper and less time-consuming means, and thus take more advantage of cyber and online only participation opportunities.

Table 8.1 Predictors of 2012 Participation

	Cyber		*Online Only*		*Online*		*Traditional*	
Party ID	−0.023	*	−0.026		−0.018		0.000	
	0.014		0.017		0.018		0.020	
Party Intensity	0.064	***	0.043	.	0.090	**	0.189	***
	0.021		0.026		0.028		0.031	
Political Knowledge	0.037	**	0.056	***	0.131	***	0.036	*
	0.012		0.014		0.015		0.017	
White	0.044		0.105		0.175	*	0.026	
	0.053		0.065		0.070		0.077	
Income	0.008		0.007		0.006		0.009	
	0.009		0.011		0.012		0.013	
Religiosity	0.026	.	0.010		−0.007		0.049	*
	0.014		0.017		0.018		0.020	
Gender	0.001		0.127	*	0.066		0.152	*
	0.044		0.053		0.057		0.063	
Intercept	0.431		0.320		0.481		0.272	
	0.106		0.129		0.139		0.153	
N	1605		1605		1605		1605	
Adj R^2	0.0171		0.017		0.0595		0.0319	

p<.05 = *, p<.01 = **, p<.001 = ***

Interestingly, party identification does not have statistically significant effects in the models. This finding is in opposition to previous findings by Williams and Gulati (2007, 2009) as well as Steinberg (2010). Their findings were likely time limited and due to the large Web 2.0 presence that Barack Obama's campaign initiated during the 2008 election cycle, and the history of Democratic candidates taking advantage of new Internet communication technologies. As these mediums are more familiar to a wider demographic of voters, the effects have subsided.

RESULTS AND ANALYSIS OF THE EFFECTS OF PARTICIPATION

Table 8.2 displays the results of probit analysis on the respondents' turnout in the 2008 general election, turnout in 2010 midterm election, and intention to turn out for the 2012 election. It should be kept in mind that the majority of the sample was ineligible to vote in 2008 due to being under 18 at the time of the election. Of those old enough, 243 respondents (15.14% of the sample) voted in 2008 and 165 (10.28% of the sample) in 2010. Therefore, most of the weight of analysis should be given to the subjects' intentions to vote in the next election.

Table 8.2 Previous Turnout and Vote Intentions

	Vote 2008		*Vote 2010*		*2012 Vote Intentions*	
Cyber Participation	−0.031		0.084		0.225	**
	0.087		0.098		0.076	
Traditional Participation	0.286	**	0.41	***	0.196	*
	0.09		0.103		0.077	
Online Participation	0.066		0.171		0.226	*
	0.129		0.16		0.103	
Party ID	−0.035		0.001		0.063	*
	0.024		0.026		0.025	
Party Intensity	0.126	**	0.117	**	0.268	***
	0.04		0.044		0.037	
Political Knowledge	0.056	*	0.052	*	0.029	
	0.022		0.025		0.019	
Race	0.511	***	0.308	**	0.404	***
	0.091		0.102		0.089	
Income	−0.039	*	0.017		0.015	
	0.017		0.019		0.015	
Gender	−0.052		0.04		−0.07	
	0.081		0.089		0.071	
Trust in Government	−0.033		−0.025		0.042	
	0.076		0.082		0.066	
Satisfaction Level	−0.163	*	−0.099		−0.08	
	0.066		0.073		0.059	
Intercept	−1.487		−2.336		−0.602	
	0.231		0.275		0.202	
N	1605		1605		1605	
Psudo R^2	0.064		0.06		0.079	

p<.05 = *, p<.01 = **, p<.001 = ***
Note: Most of the students surveyed had never voted before, and were not old enough to have voted in the 2008 elections.

In 2008 and 2010 traditional participation is the only means of engagement to be a significant predictor of turnout. This is hardly surprising given the small number of voters in those elections, the majority of which would be older non-traditional students in the sample as younger students would not have been eligible. Previous chapters have shown that older individuals are less likely to engage in cyber participation, and thus cyber participation would not be expected to be significant for older voters. However, when looking at intention to vote of the subject, cyber and online participation are both statistically significant at the $p < .05$ level and have larger effects than traditional participation. The best way to understand the actual effect of these means of participation is to look at the marginal effects.

Manipulating only the various forms of participation allows for an exploration of the effects of a person's decision to engage in each form and the effect on the likelihood that they intend to vote. For a student who engages in only traditional participation, but not online or cyber participation, the predicted value of their intention to vote is 64.16%. However, should they engage in only online participation or only cyber participation the predicted value of the intention to vote goes up to 68.24% or 65.53% respectively. A respondent who chooses to engage in none of the various forms of participation has a predicted value of intent to vote at 57.43%. This means that any form of participation raises the intention to vote from between 6.73 to 10.81%. And that engaging in online or cyber participation has the same if not more of an effect on the intention of a student to vote in an upcoming election.

Additionally, should a respondent choose to engage in all three forms of participation, the predicted value of their intention to vote rises to 80.57%. This is over 16% higher than the intention to vote given traditional participation alone. Taken together, this suggests that both online and cyber participation have a positive and significant impact on the intention of young people to turn out in future elections.

CONCLUSIONS

The focus on young people in this study allows us to think about the impact of new and evolving forms of participation on future generations of voters. If these new means of participation are eliciting a different group or event just engaging potential young voters in way previously not seen, then it is important to know how this will affect voting behavior. As we are seeing, access to the Internet allows for people to participate in politics through means that are less economically taxing or time-consuming than traditional methods. With the advancing presence of the Internet in daily life, more people are able to participate than ever before. These effects are even more potent for young citizens who have grown up using this medium. By the time a student reaches college and voting age, use of the Internet is a standard in daily life. Moreover, young people tend to be technology adopters, and this may translate towards the adoption of new means of participation.

This study has shown that online and cyber participation is frequent among potential young voters, with over 80% of students sampled becoming politically engaged using the Internet. Additionally, these forms of participation are not biased toward the socioeconomic characteristics of traditional participation. Thus implying, they are a natural means for young voters to engage in. Once engaged through the Internet or an online social networking tool, like Facebook, a young person is more likely to vote. From a civic engagement standpoint, perhaps just encouraging students in the classroom to voice their political opinions may in turn lead to broader engagement and increased political participation. Given the relatively high degree of political knowledge students seem to either already posses or are able to gain through engagement, such action would also then lead to a more informed electorate.

NOTES

1. This statistic is gathered from the survey used in this study.

2. "Contrary to the hopes of some advocates, the internet is not changing the socio-economic character of civic engagement in America. Just as in offline civic life, the well to-do and well-educated are more likely than those less well off to participate in online political activities such as emailing a government official, signing an online petition or making a political contribution" (Smith et al. 2009).

3. Questions were worded similar to the Pew Internet and American Life Survey (2008).

4. For purposes of comparison, a less nuanced version of coding for income was also used as found in Steinberg (2010). The results from both analyses were similar, and thus the more nuanced version is reported here.

REFERENCES

Berelson, Bernard R., Paul F. Lazarsfeld, and William N. McPhee. 1954. *Voting: A Study of Opinion Formation in a Presidential Campaign.* Chicago: University of Chicago Press.

Best, Samuel J., and Brian S. Krueger. 2005. "Analyzing the Representativeness of Internet Political Participation." *Political Behavior* 27 (2): 183–215.

Bimer, Bruce. 1998. "The Internet and Political Transformation: Populism, Community, and Accelerated Pluralism." *Polity* XXXI (1): 133–160.

Brady, Henry E., Sidney Verba, and Kay Lehman Schlozman. 1995. "Beyond SES: A Resource Model of Political Participation." *American Political Science Review* 89 (2): 271–294.

Bucy, Erik P. 2004. "Interactivity in Society: Locating an Elusive Concept." *The Information Society* 20: 373–383.

Denny, Kevin, and Orla Doyle. 2009. "Does Voting History Matter? Analysing Persistence in Turnout." *American Journal of Political Science* 53 (1): 17–35.

DiMaggio, Paul, Eszter Hargittai, W. Russell Neuman, and John P. Robinson. 2001. "Social Implications of the Internet." *Annual Review of Sociology* 27: 307–336.

Downs, Anthony. 1957. *An Economic Theory of Democracy.* New York: Harper & Brother.

Gibson, Rachel K., Wainer Lusoli, and Stephen Ward. 2005. "Online Participation in the UK: Testing a 'Contextualised' Model of Internet Effects." *British Journal of Politics and International Relations* 7 (4): 561–583.

Gronlund, Kimmo. 2007. "Knowing and Not Knowing: The Internet and Political Information." *Scandinavian Political Studies* 30 (3): 397–418.

Gulati, Girish J. "Jeff," and Christine B. Williams. 2007. "Closing the Gap, Raising the Bar: Candidate Web Site Communication in the 2006 Campaigns for Congress." *Social Science Computer Review* 25 (4): 443–465.

Herrnson, Paul S., Atiya Kai Stokes-Brown, and Matthew Hindman. 2007. "Campaign Politics and the Digital Divide: Constituency Characteristics, Strategic Considerations, and Candidate Internet Use in State Legislative Elections." *Political Research Quarterly* 60 (1): 31–42.

Ho, Daniel E., Kosuke Imai, Gary King, and Elizabeth A. Stuart. 2007. "Matching as Nonparametric Preprocessing for Reducing Model Dependence in Parametric Casual Inference." *Political Analysis* 15: 199–236.

Ho, Daniel, Kosuke Imai, Gary King, and Elizabeth Stuart. 2007. "Matchit: Nonparametric Preprocessing for Parametric Causal Inference." *Journal of Statistical Software.*

Howard, Philip N. 2005. "Deep Democracy, Thin Citizenship: The Impact of Digital Media in Political Campaign Strategy." *The ANNALS of the American Academy of Political and Social Science* 597 (1): 153–170.

Huckfeldt, Robert, and John Sprague. 1995. *Citizens, Politics, and Social Communication: Information and Influence in an Election Campaign.* New York: Cambridge University Press.

Jansen, Jim. 2010. *Use of the Internet in Higher-income Households.* Washington, DC: Pew Internet & American Life Project.

Jennings, M. Kent, and Vicki Zeitner. 2003. "Internet Use and Civic Engagement: A Longitudinal Analysis." *Public Opinion Quarterly* 67: 311–334.

Jensen, Jakob Linaa. 2003. "Public Spheres on the Internet: Anarchic or Government-Sponsored—A Comparison." *Scandinavian Political Studies* 26 (4): 349–374.

Jensen, Jakob Linaa. 2003. "Virtual democratic dialogue? Bringing together citizens and politicians." *Information Polity* 8: 29–47.

Jensen, Michael J., James N. Danziger, and Alladi Venkatesh. 2007. "Civil Society and Cyber Society: The Role of the Internet in Community Associations and Democratic Politics." *The Information Society* 23: 39–50.

Klofstad, Casey A. 2007. "Talk Leads to Recruitment: How Discussions about Politics and Current Events Increases Civic Participation." *Political Research Quarterly* 60 (2): 180–191.

Kobayashi, Tetsouro, Ken'ichi Ikeda, and Kakuko Miyata. 2006. "Social Capital Online: Collective use of the Internet and Reciprocity as Lubricants of Democracy." *Information, Communication & Society* 9 (5): 582–611.

Leighley, Jan E., and Jonathan Nagler. 1992. "Individual and Systemic Influences on Turnout: Who Votes? 1984." *The Journal of Politics* 54 (3): 718–740.

Mesch, Gustavo S., and Stephen Coleman. 2007. "New Media and New Voters: Young People, the Internet and the 2005 UK Election Campaign." In *Young Citizens in the Digital Age: Political Engagement, Young People and New Media*, ed. Brian D. Loader, 35–47. New York: Routledge.

Paolino, Philip, and Daron R. Shaw. 2003 "Can the Internet Help Outsider Candidates Win the Presidential Nomination?" *PS: Political Science and Politics* 36 (2): 193–197.

Putnam, Robert D. 2000. *Bowling Alone.* New York: Touchstone.

Ragsdale, Lyn, and Jerrold G. Rusk. 1993. "Who Are Nonvoters? Profiles form the 1990 Senate Elections." *American Journal of Political Science* 37 (3): 721–746.

Rosenstone, Stephen J., and John Mark Hansen. 1993. *Mobilization, Participation, and Democracy in America.* New York: Macmillan Publishing Company.

Ruxton, Graeme D. 2006. "The Unequal Variance t-test is an Underused Alternative to Student's t-test and the Mann-Whitney U test." *Behavioral Ecology* 17 (4): 688–690.

Schroeder, Stan. 2009. *Facebook Users Are Getting Older. Much Older.* July 7. http://mashable.com/2009/07/07/facebook-users-older/ (accessed 11 26, 2009).

Smith, Aaron, Kay Lehman Schlozman, Sidney Verba, and Henry Brady. 2009. *The Internet and Civic Engagement.* Washington, DC: Pew Internet & American Life Project.

Steinberg, Alan. 2010. "Is Social Media Networking Breaking Down the Barriers of Traditional Participation." *Midwest Political Science Association Annual Meeting.* Chicago.

Timpone, Richard J. 1998. "Structure, Behavior, and Voter Turnout in the United States." *American Political Science Review* 92 (1): 145–158.

Tolbert, Caroline J., and Ramona S. McNeal. 2003. "Unraveling the Effects of the Internet on Political Participation?" *Political Research Quarterly* 56 (2): 175–185.

Verba, Sidney, and Norman Nie. 1972. *Participation in America: Political Democracy and Social Equality.* New York: Harper and Row.

Verba, Sidney, Kay Lehman Schlozman, and Henry Brady. 1996. *Voice and Equality: Civic Voluntarism in American Politics* . Cambridge: Harvard University Press.

Wattenberg, Martin P. 2007. *Is Voting for Young People?* New York: Pearson Education, Inc.

Wellman, Barry, Janet Salaff, Dimitrina Dimitrova, Laura Garton, Milena Gulia, and Caroline Haythornthwaite. 1996. "Computer Networks as Social Networks: Collaborative Work, Telework, and Virtual Community." *Annual Review of Sociology* 22: 213–138.

Williams, Christine B., and Girish J. "Jeff" Gulati. 2009. "Facebook Grows Up: An Empirical Assessment of its Role in the 2008 Congressional Elections." *2009 Annual Meeting of the Midwest Political Science Association.* Chicago, IL.

Williams, Christine B., and Girish J. "Jeff" Gulati. 2007 "Social Networks in Political Campaigns: Facebook and the 2006 Midterm Elections." *2007 Annual Meeting of the American Political Science Association.* Chicago, IL.

Williams, Christine B., and Girish J. "Jeff" Gulati. 2006. "The Evolutionary Development of Campaign Web Sites: The U.S. Senate, 2000–2004." *2006 Annual Meeting of the American Political Science Association.* Philadelphia.

Williams, Christine B., and Girish J. "Jeff" Gulati. 2008. *The Political Impact of Facebook: Evidence from the 2006 Midterm Elections and 2008 Nomination Contest.* Waltham, MA: Bentley College.

Williams, Christine B., and Girish J. "Jeff" Gulati. 2008. "What is a Social Network Worth? Facebook and Vote Share in the 2008 Presidential Primaries." *2008 Annual Meeting of the American Political Science Association.* Boston.

Williams, Christine B., and Girish J. "Jeff" Gulati. 2008. "Your Money or Your Network: Indicators of Presidential Candidate Viability in the 2008 Nomination Contest." *2008 Annual Meeting of the Midwest Political Science Association.* Chicago.

Xenos, Michael, and W. Lance Bennett. 2007. "Young Voters and the Web of Politics: The Promise and Problems of Youth-Oriented Political Contest on the Web." In *Young Citizens in the Digital Age: Political Engagement, Young People and New Media*, edited by Brian D. Loader, 48–67. New York, NY: Routledge.

Zickuhr, Kathryn. 2010. *Generations Online in 2010.* Pew Internet and American Life Project.

Zukin, Cliff, Scott Keeter, Molly Andolina, Krista Jenkins, and Michael X. Delli Carpini. 2006. *A New Engagement? Political Participation, Civic Life, and the Changing American Citizen.* New York, NY: Oxford University Press, Inc.

Chapter 9

Weblogs

Revolutionary Media?

Simona Stano

"A warm place in winter and an airy one in summer, with an address, a mailbox, windows to look inside if you are in the neighbourhood and an open door if you want to enter." This is the definition of blog provided by the journalist Marino Niola, who considers it "more than just a communication system," since it is "a space for symbolic sharing characterized by its own customs, conventions, a deposited—but first created—code, and a common language"; that is, a virtual space that transcends physical boundaries, and "a metaphor of our rapidly changing present, as well as a forerunner of our future" (Niola 2008, 39 [our translation]). This very enthusiastic description is widely shared by both users and creators of blogs, as well as by a number of scholars and experts of media studies: weblogs are often presented as revolutionary tools able not only to make changes in the technological universe, but also to cause radical transformations in other spheres, first of all those of news-making and journalism. But are blogs really so "revolutionary"? To what extent and how have they innovated journalism? And is it correct to talk about the so-called citizen journalism (Bowman and Willis 2003) as completely antithetical to traditional newspapers?

First of all, it is essential to focus on the same definition of the term blog, which, according to Peter Merholz,[1] comes from the contraction of the words web, "network" (in the sense of the Internet) and log, "a detailed record of a voyage of a ship or aircraft." For this reason, it is also usually referred to as "logbook" or "diary," thus recalling the metaphor of navigation generally used for the Internet, and pointing out the predominance of the temporal dimension in this kind of publication (in weblogs, in fact, the chronological order is essential). The parallelism with the diary, however, should not mislead us: not all blogs represent a digital and public version of personal thoughts and anecdotes; on the contrary, in most cases, they take on a collective character, representing a means by which several people can share—within a space of virtual aggregation—different opinions on particular events, ideas on topics of common interest, and knowledge in different fields. In addition to personal blogs, in fact, there are many professional or organizational weblogs, which collect information about certain activities (cf. *Idearium.org*, a collaborative blog dedicated to the community of developers of interfaces, which gathers documentation on issues of common interest, reporting practical and technical details); the so-called journalist blogs, where many journalists (but not only) give voice to their opinions on current affairs and news or express their views on issues that cannot be deepened on common newspapers; the *watch blogs*, aiming at criticizing the errors underlying online news or other sources of information; the travelogs or travel blogs; the photologs (cf. Buba); the vlogs, mostly containing videos; the urban blogs, used in the field of public administration and political communication; and many others.

As Maurizio Dovigi (2003) states, the blog represents "an incubator for community navigators with common interests" (2 [our translation]), which can assume many different forms depending on the aims of its "author." It is essential to notice that, in any case, it represents a personal page with wide possibilities of customization, which is different from the amateur websites realized with common Web editors like Microsoft FrontPage or Adobe Dreamweaver for its low costs and its ease and simplicity. Hence comes the definition of blogs as unquestionably innovative tools under several aspects, and especially democracy, pluralism, openness, and freedom of expression. The blogger and expert in digital communication and culture Giuseppe Granieri (2005) refers to weblogs as a *rich system*: according to the author of *Blog generation*, blogs users do not compete to gain visibility, but continually transfer attention to other nodes, promoting a virtuous circle. "The weblog"—Granieri states—"does not pretend to be exhaustive, or to propose a defined reading experience; [...] on the contrary, it tends almost by definition to 'bring its readers outside'" (Ibid., 38 [our translation]), directing them towards other sources. Nobody reads only one blog, as it represents a single node in a "collective and hypertextual work" (Ibid.), which tends to be seen as a content system, that is to say a system in which everybody has some benefits: the author of the post gains attention and visibility; the author of the quote provides a quality input to the message and becomes visible himself; the reader can more easily find interesting contents. The working principle of this mechanism is very simple: the blogger A posts a comment, an article, or an essay on his blog; the blogger B reads it and comments on it with a post on his blog, linking the blogger A. The blogger C does the same, and so on. Very often, then, there is also a blogger T, who reconstructs the entire evolution of the discussion linking all contributions from other bloggers (A to S) and adding his own contribution.

This mechanism is not written nor codified, but spontaneously originated; nonetheless it has quickly become a standard within the community of bloggers: those who do not adhere to it are likely to remain on the margins of this system, and could even be excluded from it. With the huge success and the widespread use of Web search engines like *Google*, in fact, the links have ceased to be simple connectors, gaining a decisive value: as the Norwegian scholar Jill Walker has pointed out, they "have become the currency of the Web" (Walker 2002): the algorithm known as *Pagerank* uses the links to determine the popularity and thus the visibility of a site, thus establishing the order by which the results of the search are shown. "When I link to B, I give B a link. That link translates to a precise (though undisclosed) value in Google's PageRank and in other indexing systems like Blogdex. The link has a clearer value to B than the content of B's page has to me or to my readers. I pay B for B's content with my link" (Ibid.). Even without considering specific technical issues, therefore, the importance of hyperlinks is evident: the Web is configured as a scale-free or scale-invariant network, where not all nodes have the same importance or visibility, as this is attributed by the number of links associated with them. When a node needs to establish a new connection, it tends to do it with another one already having many links. In doing so, moreover, it helps to increase the number of its links, according to a mechanism in which "the rich gets richer while the poor gets poorer" (Ibid.).

In addition to this, one of the most acclaimed characteristics of blogs is interactivity: making a mutual and open exchange of information and opinions possible, according to a nonhierarchical organization, weblogs are praised for establishing a flux of communication with "no centres nor peripheries" (Maistrello 2004, 151), where the exchanges taking place among the different nodes, although being characterized by different degrees of visibility, have all the same opportunity to take part in the communication process in a direct way, without any kind of subordination. Considering its simplicity and low costs of utilization, anyone can create a blog and post anything on it, and anyone can have access to any blog, as the only prerequisite is to be connected to the Web. Weblogs are tools within everyone's means, capable of potentially giving voice to anyone who wants to express his thoughts. It is essential, in this sense, the possibility of posting comments: in almost all the blogs it is possible to comment on a post or to leave contributions or opinions; even when this option is not included, moreover, the user can leave a comment by sending an e-mail to the author of the site (as his or her address is generally indicated on the page) or by opening another blog to express his or her vision and thoughts, using the links to make reference to the considered post. Joi Ito has described this phenomenon as emergent democracy (Ito 2004): "Not only the citizen-blogger is no longer a passive receptor of one-way messages, but he has now access to powerful research and analytical tools, as well as means of expression to make his opinion known, and a network of contacts through which he can raise awareness (and sometimes even encourage the action) of other citizens" (in Maistrello 2004, 155 [our translation]). Weblogs can be described as real "boxes for ideas" (Cavedoni 2002), boxes that are not hermetically sealed, but, on the contrary, seem more open than ever to the exchange of materials, in all directions.

The gap between these new media and the most common mass media—mainly based on a unilateral communication between a transmitter, on the one hand, and a mass of more or less passively receiving addressees, on the other hand—appears therefore in many ways unbridgeable. It is in this sense that blogs represent, according to several scholars, a "new venture for journalism" (Granieri 2005, 119 [our translation]), as anyone, becoming "editor of himself" (Granieri 2004), can now assume the role of the reporter, without being necessarily required to subscribe to any journalist association or to have been trained in this field. It is the *grass root journalism*, or journalism "coming from the common people," whose exemplar case could be *SlashDot*, a blog created in 1997 by Rob Malda where it is possible to find news about various topics. Here the posts can be posted by all the users of the site, subject to the approval of the editorial team, in order to prevent the risk of *hoaxes*, which is very high as there is information coming from the most varied sources. All posts, then, can be commented on by anyone, which gives rise to long threads of discussion around them. In other cases, instead, single individuals create their personal blogs for information purposes, without undergoing filters and editorial endorsements of any kind. In this way a sort of amateur journalism (or participatory journalism, as defined by Lasica 2003) has emerged, giving people who have something to say the possibility to do it, and so making them act, in fact, as journalists, without the need of newspapers or editorial teams agreeing to publish their articles, or of institutions certifying their professional skills (cf. Staglianò 2002). According to Riccardo Staglianò, journalist of "la Repubblica," it is in this sense that we can speak about the "blog revolution," which is a revolution that finally realises the "ancient promise by the network to provide a platform to enable an unmatched, never existed before chance for everyone to real express his/her thoughts" (Ibid., 179 [our translation]). For this same reason, moreover, *Blogger.com*, one of the leading platforms for creating and managing blogs, refers to itself as the "one-button publishing for all the people": the information provided by blogs arises outside the newsrooms, directly from people who, despite never having practiced journalism, are characterised by a certain degree of "journalistic spirit." Moreover, it is a sort of alternative information, as it is published in full freedom, without the limits or censorships due to the interests or positions of the groups of power generally related to the printed newspapers.

Blogs seem to represent a "new adventure" for the so-called fourth power even for another reason, since they constitute a sort of controller always ready to re-examine the work of more or less famous journalists, promptly pointing out errors and inaccuracies. Consider, for example, the blog of Paolo Attivissimo, *Il Disinformatico*, where the author, who calls himself a "web journalist and hoaxes hunter," has pointed out the most striking cases of mistakes and imprecisions appeared on some of the most famous newspapers.

A large number of scholars and experts, therefore, agree to celebrate the incredible revolutionary capability of new media, praising its "democratic, pluralist and libertarian character" (Uboldi 2004, 148 [our translation]). By contrast, as others support, these characteristics do not constitute an absolute newness introduced by blogs, but rather seem to be intrinsic to the wider path of evolution of the Web itself. According to Giuseppe Granieri (2005), "definitely weblogs are not a new thing, at least technologically and logically speaking" (25 [our translation]). Technologically, they represent just one of the different novelties brought by the evolution of the Network, and more specifically of the *Content Management System* (CMS), which are systems for content management designed to facilitate the organization and publication of information on the Internet or corporate networks. Without going into specific technical issues, it can be said that the weblogs represent the simplest model of CMS, thanks to which it is possible to publish documents on the Web without having specific technical skills (but to type texts using a keyboard). They constitute, therefore, "an answer to a problem arisen with the Web and with the vision of its inventor Tim Berners- Lee, who wanted to make it a place where publishing contents was as easy as consulting them" (Ibid., 26 [our translation]). They don't work according to schemes that are entirely new and unrelated to the past, but through the already existing mechanisms of the content management systems. In this sense, "within the framework of the short history of the Web, blogs [...] undoubtedly represent a very significant and innovative step, but not a revolution" (De Biase 2003 [our translation]). With respect to the "logical" dimension mentioned by Granieri, despite many people believing that journalistic blogs are intended to subvert—or even to cause the death of—the traditional journalism, it should be noted that, although "on the technical side, the work of the journalist and that of the bloggers have the same content, that is information, [...] it is not possible to make a comparison" (Granieri 2004 [our translation]) between them. According to the author of *Blog Notes*, not only in a blog time, modalities of utilization and technical factors change, but also, and above all, the blogger "represents himself and only himself," he is a sort of "columnist" who freely shares his impressions of the world. The most common complaint about new media, therefore, concerns the poor reliability of the information they provide, due partly to the high degree of autoreferentiality that characterizes them and partly to the alleged lack of specific competences of those who manage them. On the one hand, it is true that the access to the world of information is increasing and that a large group of amateur journalists can "finally express themselves without imploring publishers to get published" (Dovigi 2006, 33 [our translation]); on the other hand, it is also undeniable that professional ethics and news reliability become in this way more difficult to control.

However, even the traditional print journalism does not seem to be immune to mistakes or hoaxes (which are often denounced by bloggers themselves on their pages), so it is necessary to make distinctions based not on the dichotomy "professional journalists" versus "bloggers," but rather according to the effective degree of reliability of the published contents: "When a blogger writes up daily accounts of an international conference, [. . .] that is journalism. When a magazine reporter repurposes a press release without checking facts or talking to additional sources, that is not. When a blogger interviews an author about his new book, that is journalism. When an opinion columnist manipulates facts in order to create a false impression, that is not. When a blogger searches the existing record of fact and discovers that a public figure's claim is untrue, that is journalism. When a reporter repeats a politician's assertions without verifying whether they are true, that is not" (Blood 2004; cf. Blood 2003).

Both blog and printed newspapers should pursue neutrality and objectivity, basing the information they provide on reliable sources. Only in this way, both of them may fall into the category of "journalism."

The relationship between common journalism and blogs, ultimately, is not of an either/or type. By contrast, it presupposes forms of interaction and integration that are often underestimated: since 2001, when Claudio Sabelli Fioretti's blog and *Wittgenstein* by Luca Sofri appeared, for example, the number of Italian professional journalists who have decided to experience the formula of the weblog as a tool to directly reach readers and enhance the dissemination of their opinions has increased more and more. "Listen to your readers through a weblog is a good way to improve your relationship with the public, [...] and it can have immediate returns also in your daily work" (Uboldi 2004, 159 [our translation]): a growing number of journalists working for the print press, as well as the same newspapers editorial teams, understanding the importance of blogs, have decided to promote them within their own Web pages. Incorporating the weblog of Sofri on its website, on October 15, 2002, "Il Foglio" was the first Italian newspaper to move in this direction, followed by "il Riformista," "Internazionale," and *Repubblica.it*, (which currently has more fifteen blogs, still including the initial ones opened in September 2003: Ernesto Assante's *MediaTrek*, Leonardo Coen's *Blog Trotter*, Maurizio Crosetti's *Rimbalzi*, and Federico Rampini's *Estremo Occidente*).

Many editorial teams have also created specific systems of classification for the blogs managed by their journalists: the *New York Times*, for example, has different categories for its weblogs dealing with *News and Politics*, *Business and Finance*, *Technology*, *Culture and Media*, *Health, Family and Education*, *Style, Travels and Leisure*, *Sports*, and *Opinion*. The advantages deriving from such encounters are anything but secondary: first of all, journalists have the opportunity to "know their audience, [...] but also to learn from their audience" (Uboldi 2004, 172 [our translation]), allowing people to give them feedback. Moreover, information can be combined with the expression of the same author's personality and thoughts, which are elements that generally cannot be included in the pages of common newspapers, thus increasing the readers' reliance on what they read and creating a sort of attachment to the writer. Finally, the ability to update information in real time and the possibility to cover a topic for a longer period, without the limitations imposed by conventional standards of newsworthiness, constitute a great gain too.

Beyond a priori and arbitrary positions, therefore, traditional journalism and blogs do not represent two antagonistic extremes, but rather two models that can be integrated and are already partially integrated. Moreover, the proximity between these two models is not limited to the factors discussed so far, but also embraces forms of communication far more distant in time and apparently totally disconnected from the current information blogs. A specific example can help us to better understand this aspect: *L'Osservatore Veneto* by Gasparo Gozzi (1713–1786), one of the most brilliant achievements of the Italian press of the late eighteenth century.

Belonging to the Venetian journalism of the 1760s, which was characterized by the presence of various currents and genres thanks to which the great tradition of the Venetian press was able to cope with the general crisis of the print press industry of that time, the work by Gozzi formed part of the so-called *giornali morali* ("moral newspapers"). Moving from the example of "The Spectator" by Joseph Addison and Richard Steele (1711), these newspapers placed at their center the observation and critique of the society and customs of the time, with a clear moralizing aim. Last of three editorial attempts by the author, published from February 1761 to August 1762 after the previous adventures of the *Mondo Morale* and the *Gazzetta veneta*, *L'Osservatore veneto* came to seal the Gozzian work in this field, as it represented "the literary form that better than any other previous one responded to his intention and his skills as a writer, as well as to the tastes and trends of his time" (Spagni in Gozzi 1761–1762 [1900], V [our translation]).

The result was a varied and multifaceted work, which ranged from literature to manners and morality, whose breadth of contents was due, on the one hand, to the same journalist's intentions ("I can hardly reduce jumping out from a topic to another. When I undertake a venture which is not going to end soon, it seems impossible to me to be able to get to its bottom,"[2] Gozzi 1761 [1900], 68 [our translation]) and, on the other hand, to the tastes of his audience ("it is like a collection of paintings, in which different attitudes, colours, and manners are portrayed, so that everyone can stop in front of the canvas he likes the most,"[3] Ibid., 14 [our translation]). With this respect, the numerous letters published in the magazine played a key role: the correspondence, which was already present in the *Gazzetta*, became incredibly intense in the *Osservatore*, so as to include not only comments and inquiries, but also short tales or other texts sent by friends and other senders. Thus the object of discussion became very variable and the relationship between the "observer"[4] and his correspondents (which might be called, in more contemporary terms, "interactivity") became an essential element as it represented the structure of the magazine itself.

Moreover, it should be remembered that Gozzi's model was "The Spectator," even if the difference between the English and the Italian reality caused the evolution of the Venetian newspaper in a direction quite different from that of Addison's newspaper: while in the pages of "The Spectator" the letters from the readers contained reflections and proposals full of the distinctive British practical sense, the writings received by the Venetian scholar were characterized by long and intricate rumours and by the dogmatic "philosophism" typical of those years. Gozzi soon realized that such topics, and especially the intricate and severe form that characterized them, did not suit the nature and purpose of the newspaper he had in mind.

Many of these correspondents, then, according to the custom of the time, thought they had to repay the hospitality granted to them in the magazine with praise and compliments to the author, which, instead, offended his sincere modesty and even made him fear that readers might suspect the letters were fictitious. These and other factors convinced Gasparo Gozzi of the need of reviewing all the letters received. Therefore he asked some other scholars from the Granelleschi Academy to join him in the editing of the newspaper, whose title was changed in "Gli Osservatori veneti." With this name it survived until August 1762.

Comparing this work with some of the most read contemporary journalistic blogs gives evidence that, beyond the obvious differences related to production and distribution technologies, or the linguistic and stylistic standards of writing, there is a strict connection between these models. This connection, despite being seemingly invisible, is in fact very strong, as three different aspects point out.

The first one is interactivity that is that unavoidable interaction between author and audience which seems to offer writers a starting point for new considerations, thus determining the same structure of the published contents. Both for the *Osservatore* and many blogs, in fact, the relationship with the public is essential and defines the topic of discussion. We have already discussed the centrality of the interactive relationship between bloggers and the public; with respect to Gozzi's newspaper, it should be noticed that the letters from the correspondents addressed the discussion to a direction rather than to another, providing insights for the author's considerations or sometimes even the same texts he decided to publish as he received them. In many cases, indeed, Gozzi received tales or opinions written by his readers (the most evident case is that of the so-called painter of moral portraits, whose "portraits" of the human soul found space in several issues of the magazine), while sometimes people asked him to cover certain topics or offered him opinions on literary or political issues. By contrast, it is undeniable that there is a substantial difference between the kind of interactivity offered to the public by the *Osservatore* and by blogs, which is mainly related to the temporal dimension: while the modalities of correspondence with Gozzi and the production procedures of the magazine caused some delay in the communication process between the readers and the journalist, weblogs are characterized by instantaneity, which seems to permit the full realization of the principle of interactivity. Nevertheless, if the term interactivity is conceived, according to its common usage, as the ability to act in correlation or reciprocity with others, and if we consider that the same structure of the *Osservatore* was based on the correspondence, we can easily conclude that the presumed revolution introduced by the weblogs is not a "revolution" per se, but rather the development (or, if anything, the sublimation) of a feature that already existed, albeit in a different form, in other and antecedent media. The new medium represents the achievement of a degree of interaction between different nodes which had never been reached before, but this does not mean that interactivity arose just with blogs. As proved by the centrality of the relationship between the author and his readers in the Gozzian work, the presence of "interactive forms" was already central for previous means of information and communication. This is the first reason why the term "revolution" does not seem so appropriate to describe the innovations introduced by blogs.

A second element of similarity between the Venetian newspaper and the several journalistic blogs that populate our daily "mediascapes" (Appadurai 1990, 9) is the openness of their contents, which is the result of the critical sensibility of the authors, who are able to keep together different topics and fields in order to grasp and discuss the problems of contemporary reality. Although Gozzi is more focused on the arts and literature, while Sofri, Coen, Assante, and colleagues generally seem more interested in politics and current events, we can perceive in any case a great ability to catch different aspects of society, which is an attitude that leads the writers to take a careful and critical look upon the events they describe. Both the Venetian writer and bloggers, therefore, could be described as accurate "observers" of reality, whose purpose is to provide a multi-perspective portrait of the human world, capable of embracing all the many nuances that colour it. Consider, for example, the comparison between the eighteenth-century magazine and the blog of Luca Sofri: Gozzi explicitly declares his goal by choosing the title *L'Osservatore veneto* (lit. *The Venetian Observer*), with immediate reference to Addison and Steele's "The Spectator" and so to the attempt of establishing himself as an "observer" of reality. This same idea, then, is repeated several times in his writings, which consider multiple topics. And Sofri shows the same tendency through a *tagline*, consisting of a quote from the famous cartoonist Schulz, "It rains on the just and on the unjust," followed by the comment "And what should we do here in the middle?" (www.wittgenstein.it/.). As the same journalist explained in an interview (2006), these words express "the intention of going beyond a black and white vision, trying to grasp all the different nuances that colour reality" [our translation]. Such an intention finds expression in the variety typical of his posts (and of almost all the texts posted on this type of blogs), which embrace various topics, such as current events, culture, politics, entertainment, art, technology, and sometimes even science and sports.

Finally, it is necessary to consider the moralistic intent that relates Gozzi's journal to contemporary journalistic "diaries": observation and discussion are never ends to themselves, but rather aim at orienting, more or less directly, the readers' thoughts and actions. This is more evident in the *Osservatore*, as Gozzi explicitly clarifies that he aims at "portraying" human vices and virtues through various devices, like the numerous stories and allegorical tales, the moral reflections, or the many letters from the above mentioned painter of moral portraits. In the tale of the mouse and the pike (No. XLIII, July 1, 1761), for example, the writer tells us the story of two animals that, trying to overcome the boundaries of their safe and known world (the basket in the barn for the mouse, the Piave river for the pike), come across an unknown and dangerous dimension, inhabited by other animals which are "bigger" and stronger than them. Thus they experience sometimes an insurmountable envy (as in the case of the mouse, which would have liked to be as other animals that are loved by men),[5] sometimes even extreme dangers that can lead them to death (as in the case of the cat eating the mouse, and the "greater and stronger" fish eating the pike). At the end of these tales, Gozzi himself clarifies their meaning, stating that, despite being works of fiction, they can be interpreted as metaphors of the same human life:

> Oh le son favole! Egli è il vero. Ma, se in iscambio di topi e lucci io volessi mettere o Ambrogi o Piergiovanni o altro, egli si vedrebbe che alcuni, essendo usciti per altezza di ingegno fuori delle ceste o de' rigagnoli per correre e nuotare in più largo spazio, non hanno mai avuto un bene al mondo (Gozzi 1761, no. XLIII, July 1).

And these are but two examples of the many short stories published by the author in almost two years of publishing the magazine, which was also characterized by the presence of numerous notional considerations on various moral issues (such as reasoning about the excessive self-love—no. VIII, February 28, 1761; the analysis of the difficulty of judging men—no. X, March 7, 1761; and many others). As mentioned before, moreover, it is essential the role played by the letters of the painter of moral portraits, who introduces himself in the issue of February 14, 1761, outlining the purpose of his correspondence, that is to share with the author and the other readers his works, consisting not in canvas and pictures, but rather in verbal portraits of "attitudes and manners":

> Mio Signore. Il buon sapore della pittura s'è così ampiamente allargato, che ogni casa è provveduta di qualche opera mirabile di questo genere. . . . Io non sono né pittore, né ricco uomo da potere provvedermi dell'opere . . .; sicché non avendo né danari da spendere, né attività per adoperare i colori, ho pensato un nuovo modo da guarnire un piccolo stanzino, come posso. . . . In quelle poche ore che m'avanzano dall'altre mie occupazioni, con certi ferruzzi lavoro alcune cornici; onde in non so quanti mesi m'è riuscito di scrivere in alquante cartapecore, che poi le vo incorniciando e facendone piccioli quadri, certe figure, non di visi o di corpi, ma *d'animi e di costumi;* . . . Vi mando la copia di alcuni d'essi ritratti, acciocché vediate la maniera del mio dipingere senza pennello (L'Osservatore veneto 1761).

His intent is explicit and defined: he aims at providing some portraits of men, not by painting their faces and their bodies with a brush, but drawing their minds and their costumes with his pen. So he proposes to Gozzi a series of "moral" portraits, which build on descriptions of particular traits of a man (precisely as a common visual portrait) to finally come to more abstract moral precepts on humanity as a whole. In the *First Portrait* ("Ritratto Primo"), published in the issue no. IV, February 14, 1761, for example, the description of Lysander, a man praised by all people for his kindness but in fact hostile to the visits of friends, represents a way to introduce a moral principle that invites us not to confuse reality and appearance.

The case of nowadays bloggers, however, is somehow different, since the criticism and irony of their posts make them look as "controllers" of public affairs, always ready to bring to light the weaknesses of the political and socioeconomic world, to denounce them openly, thus making people aware of them. Nonetheless, they are moved by the same moralistic intent we can find in the Gozzian work, as in many cases they try to "portray," analyze, or comment on what they consider the problems of the society we live in, thus making people become aware of them and stimulating their response. Both in the case of the *Osservatore* and that of *Wittgenstein*, *Estremo Occidente*, *The Caucus*, *City Room*, and many other contemporary digital "diaries," information is always tied to opinions, and the same receivers are called to think over the proposed topics.

These and other minor similarities clearly point out that, despite being considered as revolutionary and absolutely innovative media, weblogs are characterized by a sort of historical "skeleton" whose constitutive elements were already present in other antecedent media. There is no doubt that blogs have brought important innovations in the world of information and communication, and that something has changed and is currently changing, but the presupposed "revolution" they are commonly associated with is ostensible, at least if the term revolution is conceived, according to its original meaning, as a total subverting of previously existing settings or models. Too often, especially in fields such as technology, characterized by constant discoveries and newness, there is a tendency to hide, or at least not to consider the elements of continuity with the past, as if they could cause a loss of importance of the new inventions and discoveries. The "new," the "original," the "revolutionary" innovation is generally exalted to the extreme, drawing the attention more to the differences than to the similarities with the past. As Sergio Maistrello (2004) stated, "the blog itself is just a tool, and even not so innovative, to be honest. But it is a tool that is giving power to people [...]. It is too early to say whether it will be a revolution or not: revolutions—to use a term that is often abused in computer sciences—are not made by tools, but by the people who use these tools to improve themselves and the society they live in" (156 [our translation]).

If we want to speak about a revolution, therefore, we should not look at blogs and their characteristics—whose novelty, as we have analyzed, is somehow partial—but to our way of using them. And in order to make such use good and effective, we must get a better knowledge of the tools we are using, turning our attention not only to the future but also, and especially, to the past. It is necessary, in other words, to learn the lesson of the French philosopher Bernard of Chartres (twelfth century): "We are like dwarfs on the shoulders of giants, so that we can see more than them, and things at a greater distance, not by virtue of any sharpness of sight on our part, or any physical distinction, but because we are carried high and raised up by their giant size." Forgetting about those "carrying" us high, thus making us see "more than them," does not make our enterprise of knowledge bigger; on the contrary, it prevents us to move forward. Only trying to overcome the claim according to which blogs are unique tools totally disconnected from traditional media, we could make them better means of information and communication, therefore really turning them into "revolutionary" media.

NOTES

1. Who, in 1999, was the first to use the expression "we blog" in one of his posts, inventing the verb *to blog*.

2. "Il balzare da un argomento all'altro mi riesce di stento minore. Quando piglio per le mani un'impresa che non sia per finir tosto, mi pare impossibile di poterne mai venire a fine" (*L'Osservatore veneto* XV, March 25, 1761).

3. "è come una raccolta di pitture, in cui diversi atteggiamenti, coloriti e maniere si veggono; sicché ognuno possa arrestarsi dinanzi a quella tela che più gli piace" (*L'Osservatore Veneto*, no. III, February 11, 1761).

4. Which is the English word for *osservatore*.

5. "A poco a poco incominciò ad avere un travaglio che non avea provato nella sua casetta; imperciocchè comparando sé medesimo ad altri animali vezzeggiati dagli uomini, o maggiori di sé, veniva roso da un tarlo continuo d'invidia, e avrebbe voluto uguagliarsi ora a questo, ora a quello" (*L'Osservatore veneto*, no. XLIII, July 1, 1761).

REFERENCES

Appadurai, Arjun. 1990. "Disjuncture and Difference in the Global Cultural Economy," *Public Culture* 2, 1–24.

Blood, Rebecca. 2004. *A few Thoughts on Journalism and What Can Weblogs Do About It.* Accessed September 12, 2013. http://www.rebeccablood.net/essays/what_is_journalism.html.

Blood, Rebecca. 2003. *Weblog. Il tuo diario online*. Milan: Mondadori.

Bowman, Shayne and Chris Willis. 2003. *We Media: How Audiences Are Shaping the Future of News and Information*. http://www.hypergene.net/wemedia/weblog.php.

Cavedoni, Antonio. 2002. *Weblog, una scatola per le idee*. http://cavedoni.com/2002/10/weblog.

Colombo, Fausto. 2000. *L'informazione nei tempi di Internet.* Accessed September 12, 2013. http://www.mediamente.rai.it/biblioteca/biblio.asp?id=461&tab=int&tem=47.

De Biase, Luca. 2003. *La rivoluzione esagerata dei blog.* Accessed September 12, 2013. http://www.debiase.com/storie_Contenuto.asp?Cod_Argomento=22&Cod_Storia=260&ADDR_CALL=A.

Dovigi, Maruizio. 2003. *Weblog Personal Publishing*. Milan: Apogeo.

Dunlop, Tim. 2013. *If You Build It They Will Come. Blogging and the New Citizenship*. Accessed September 12, 2013. http://evatt.org.au/papers/if-you-build-it-they-will-come.html.

Glaser, Mark. 2004. *"Watchblogs" Put the Political Press Under the Microscope.* Accessed September 12, 2013. http://ojr.org/ojr/glaser/1076465317.php.

Granieri, Giuseppe. 2005. *Blog Generation.* Bari: Laterza.

Granieri, Giuseppe. 2004. *Blog FAQ 1.0.* Accessed September 12, 2013. http://www.bookcafe.net/blog/blog.cfm?id=146.

Ito, Joi. 2004. *Weblogs and Emergent Democracy.* Accessed September 12, 2013. http://joi.ito.com/static/emergentdemocracy.html.

Lasica, Joseph D. 2003. *Participatory Journalism Puts the Reader in the Driver's Seat.* Accessed September 12, 2013. http://www.jdlasica.com/2003/08/07/what-is-participatory-journalism/.

Maistrello, Sergio. 2004. *Come si fa un blog.* Milan: Tecniche Nuove.

Negroponte, Nicholas. 1995. *Being Digital.* New York: Vintage Books.

Niola, Marino. 2008. "Blog, il nuovo villaggio più o meno globale, tra luogo e comunità," *La Repubblica*, July 29.

Rifkin, Jeremy. 2000. *L'era dell'accesso.* Milan: Mondadori.

Rosen, Jay. 2004. *Brain Food for BloggerCon: Journalism and Weblogging in Their Corrected Fullness.* Accessed September 12, 2013. http://journalism.nyu.edu/pubzone/weblogs/pressthink/2004/04/16/con_prelude.html.

Rosen, Jay. 2004. *The Weblog: An Extremely Democratic Form in Journalism.* Accessed September 12, 2013. http://journalism.nyu.edu/pubzone/weblogs/pressthink/2004/03/08/weblog_demos.html.

Staglianò, Riccardo. 2002. *Giornalismo 2.0.* Rome: Carocci.

Uboldi, Francesco. 2004. "Il blog in redazione. Tra news, diari e commenti, la prima interattività giornalistica." *Problemi dell'informazione* 2, 145–175.

Walker, Jill. 2002. *Links and Power: The Political Economy of Linking on the Web.* Accessed September 12, 2013. http://cmc.uib.no/jill/txt/linksandpower.html

Chapter 10

The Transformation of Leisure in the Digital Age

Ana Viñals Blanco

Information and communication technologies (from now on, ICTs), Internet, and the relentless development of the so-called social media (virtual social networks, blogs, wikis, virtual communities, etc.) have notably influenced various aspects of daily life and everyday habits. As a result, it has not been possible for leisure, the field of analysis for this article, and understood here "as an integral human experience, that is, a total complex experience focused on desired, autotelic and personal actions" (Cuenca 2006, 14), to be left on the sidelines of this digitalization.

The University of Deusto Leisure Studies Institute is a national center of reference that has been working on research into "experiential leisure" from a multidisciplinary perspective (Cuenca, Aguilar, and Ortega 2010, 44–55) for twenty-five years. Based on this institute's own humanist philosophy, the sphere of leisure belonging to a postmodern and digital "network society" (Castells 2006) will now be analyzed. This is a society in which prevail the abundance of data (Reig 2012), postmaterialist values (referring to self-expression and personal fulfilment), consumerism (Bauman 2002), and the unstoppable growth of ICT, Internet, and social media. This is a digital and information society which has brought about the digitalization of different social spheres and everyday habits, such as communicating, socializing, shopping, playing, doing sports, traveling, reading, or visiting a museum, at breakneck speed. In a nutshell, an era that has inevitably brought about a transformation in the way of experiencing leisure.

However, being aware of ICT influence on the transformation of leisure styles, at the same time it is important to stress that technology has not only served as an aid to traditional leisure products and services, but has also promoted the emergence of new essentially digital leisure activities with their own characteristics making them different. The influence of technology on the world of leisure has led us towards a digitalization of the leisure experience or, in other words, towards "a new paradigm of experience" (Cuenca 2008), which on this occasion we catalogue as "digital leisure."

In this context, the purpose of this article is to provide a critical reflection on leisure in the twenty-first century; a reflection centered on the need to foster, with prior education, digital leisure practices that lead to learning and human development. This is a form of leisure that is ultimately developed under the guidelines of experiential, humanist, and positive leisure which will now be described. Therefore, the aim of this article is to analyze this emergent digital leisure, expanding upon the characteristics attributed to it.

The article is structured into three main sections. First of all, the present-day leisure phenomenon is analyzed, explaining its importance and the concept of leisure we begin with. Secondly, the influence had on leisure by the values belonging to a post-materialist, consumer, and "liquid" society (Bauman 2002). In third place, basing ourselves on the previously mentioned theory, digital age leisure is described, that is, "digital leisure," "e-leisure," and "online leisure," among others; a way of understanding leisure which not only includes leisure practices that have undergone the "technologization" of traditional leisure, but also a type of leisure that, in turn, has perceived the digitalization of the leisure experience as a whole and the emergence of new types of leisure related to the use of Internet. Finally, we provide a reflection which, by way of summing up, pays special attention to the need to have positive and healthy digital leisure experiences. This is a kind of leisure that undoubtedly requires prior education and/or training in digital skills and competences.

Lastly, it should be pointed out that this article is not conclusive, due to the fact that we are dealing with a phenomenon that is a live issue of a changing nature. What is intended is to present the key factors of a reality that is both popular and complex, and which once again deserves a rigorous scientific study from the humanist and experiential perspective of understanding leisure, which is the starting point here.

THE PHENOMENON OF LEISURE TODAY: TOWARDS EXPERIENTIAL LEISURE

There are numerous indicators pointing at the current relevance of the leisure phenomenon. However, there are two main features that stand out and are clear proof of the important role that leisure plays in today's society. On the one hand, global statistics related to various spheres of leisure, such as tourism (World Trade Organization [WTO]), sport (The Fédération Internationale de Football Association [FIFA]), culture (United Nations Educational, Scientific and Cultural Organization [UNESCO]) and recreation (Groupe d'Etude des Maladies de l'Oreille [GEMO]), are a clear signal that leisure not only constitutes a powerful factor in economic and social development, but its activities also dominate contemporary societies. On the other hand, the fifth wave of the World Values Survey (2005–2008), performed by a network of social scientists from all over the world, highlights the fact in the "Perceptions of Life" section that 74.5% of the people surveyed consider leisure/free time to be quite or very important. This is evidence of the importance of leisure as regards individual valuations.

It can therefore be claimed that leisure as a value in itself has become an important pillar of human development in the twenty-first century (Cuenca 2010). Furthermore, whereas it is attributed a residual character in Solid Modernity, understood as free time devoted to recovery from work-related fatigue, it is currently understood and valued as a "true private, free, enjoyable and autotelic experience which the consumer market takes charge of supplying" (Cuenca 2008, 17). In other words, leisure has gone from being considered a harmful counter-value to a primary value: citizens as a whole value the free time they have in a special way as opposed to work time. This is the reason why, in these times of crisis in finance, economy, and values, lack of employment is such a worrying and burning issue. Yet, how is this concept of leisure generally perceived? Which concept of leisure is our starting point?

Over the last few decades much work and research and many studies on leisure and its meaning have been produced (see, for example: Csikszentmihalyi and Douglas 1991; Cuenca 1993, 1995, 2000, 2003, 2004, 2007; Degraaf, Jordan and Degraaf, 1999: 2–4; Edginton 1995; Godbey 1990; Goodale and Godbey 1988; Haywood, Kew, and Bramham 1990; Kelly and Freysinger 1999, 14–17; Kraus 2001; Leitner et al. 1989; Neulinger 1981; Torkildsen 1986; Veal 1987; Wilson 1988). Leisure has thus gone from being related to availability of free time, idleness, or the carrying out of an activity (as it has so often been identified), to being understood as a satisfying experience. It is therefore a social phenomenon that evolved from fact to experience (del Valle 2011).

Therefore, in accordance with the "International Charter for Leisure Education," adopted by the World Leisure and Recreation Association (WLRA) in 1993, we consider leisure to be:

1. A specific area of human experience with its own benefits, among which are the freedom of choice, creativity, satisfaction, enjoyment, and greater happiness. It includes broad forms of expression or activity whose elements are often not only of a physical nature but also intellectual, social, artistic, or spiritual.
2. An important resource for personal, social, and economic development and an important feature of quality of life. Leisure is also a cultural industry that creates employment, goods, and services.

3. A form of leisure that promotes general good health and well-being by offering a variety of opportunities that enable people and groups to choose activities and experiences that meet their own needs, interests, and preferences.
4. A fundamental human right like education, work, and health which nobody should be deprived of for reasons of gender, sexual orientation, age, race, religion, beliefs, standard of health, disability, or economic status.

In short, this article's starting point is that of the notion of leisure as an experience, as a resource for development, a source of health and prevention of physical and mental illnesses, a human right based on having basic living conditions covered, a symbol of quality of life, and a possible economic potential. In the words of Dr. Manuel Cuenca Cabeza, founder of the University of Deusto Leisure Studies Institute, we understand the leisure experience as being: "a type of human experience which the subject perceives in a satisfactory, but non-obligatory and non-essential, way (…) A free, satisfying human experience with an end in itself; that is, voluntary and separate from need, understood as a primary need" (Cuenca 2004, 45)

Nevertheless, it is important to stress that the current society has turned leisure into a synonym of consumerism and therefore tends to be objectively materialized through the practice of leisure activities. Thus, on this occasion, we insist that here we understand leisure from its subjective side, that is to say, from what the practice of leisure means for the person him/herself. Ultimately, leisure is doing what you feel good about, a rewarding experience had because one wants to, and nobody, or nothing, can force you to do it. However, the analysis starts out with the idea that it is important to detect the most popular and valued social practices of leisure in society, although this might not be the best means of understanding leisure as regards individual living experiences.

In addition, research into the leisure phenomenon also indicates that it is not enough to live a Leisure Experience, but what is more, every endeavour should be made to make it a "valuable human experience of leisure" (Cuenca 2010, 90). Leisure time should be used to become personally involved, provide ideas, acquire skills, generate knowledge, live new experiences, develop ourselves in a creative way or help others. These are Valuable Leisure Experiences (VLEs) that require previous training and education in leisure, in which the values, attitudes, and skills needed to experience a mature type of leisure can be learn; a type of leisure that demands commitment, participation, motivation, and effort.

The author therefore describes four kinds of leisure which, in one way or another, lead us to enjoy more or less valuable leisure experiences: Active Leisure, Substantial Leisure, Creative Leisure, and Dignified Leisure.

Active Leisure is identified with activities that make the person get involved and provide something themselves. This is a type of leisure chosen, desired, and wanted by the person experiencing it and therefore encompasses activities involving both physical and mental activities. It is not a form of leisure that involves movement in the most direct sense of the word, but rather implies a proactive attitude. On the other hand, Substantial Leisure is defined as that which involves leisure practices carried out repeatedly over time in the systematic pursuit of intense satisfactions. According to Robert A. Stebbins, who calls this substantial leisure "serious leisure," these are "systematic pursuit of an amateur, a hobbyist, or a volunteer activity sufficiently substantial and interesting for the participant to find a career there in the acquisition and expression of a combination of its special skills, knowledge, and experience" (Stebbins 1997, 17). This type of leisure is characterized by perseverance over time, continuous effort, training, lasting benefits, identification, social interaction, sense of belonging to a social world, and free of charge.

Thirdly, we have leisure practices carried out from a perspective based on creative values, understanding the notion of creativity as the "ability to spontaneously change the way in which we contemplate, think or act in the world" (Csikszentmihalyi 1996). Creative Leisure's environment for development is more cultural, an area which is characterized by separating itself from instinct and ease where the person acquires a more complex vision and enjoyment of the world, but one that is much more enriching, satisfying, and human in the long run (Cuenca 2010, 96). Here, creative experiences related to creation and re-creation are lived, experiences that need a large dose of motivation (self-motivated), awareness, and activity. The fourth and final type is that which refers to the so-called Dignified Leisure, that is, the type of leisure in greatest demand in the twenty-first century which includes solidarity, sustainable and inclusive leisure. These types are basically motivated by the satisfaction obtained by helping others by means of solidarity initiatives; in short, experiences in active, substantial, creative and dignified leisure that give meaning to our lives.

In conclusion, leisure today is conceived as a central element in the hierarchy of citizens' values, a kind of leisure that provides numerous (psychological, sociological, and economic) benefits at both an individual and social level. For this reason, the philosophy on which we are basing ourselves here is a positive and human leisure in which the person and the living of his/her experiences are to be found at the heart of the action, regardless of the activity being performed. Therefore, we understand leisure as a source of satisfaction and happiness; and that is why, today more than ever, the popular phrase: "we do not live to work, but work to enjoy leisure" (Monteagudo 2008, 9) is of great relevance. Of course, in the last few decades, the sphere of leisure has been affected by the development of ICT and cyberspace, and a social context based on postmodern values. We therefore find ourselves facing what is a clear digitalization of leisure experiences; a new paradigm of the experience (Cuenca, 2008) which adds specific characteristics to the practices and experiences of leisure. Leisure practices belonging to a digital and post-modern age are thus being designed.

LEISURE IN THE DIGITAL AGE

As previously mentioned, leisure has inevitably been affected by the influence of factors belonging to a digital age and a society that is centred on networks and connections. Not only have leisure activities changed but also, more importantly, the way of practising and enjoying them. External factors deriving from a postmodern society founded on "here and now," constant change, the development of virtual space, the relentless impulse of a whole framework of technological tools that are cheaper and easy to use, along with the democratization of broadband Internet use, have all resulted in the generation and impulse of a kind of leisure that responds to the prevailing model: digital leisure.

In a large share of the research undertaken into the leisure phenomenon, it is fundamental to take into account the variables of time and place. Therefore, it is vital to consider the fact that, with a digital age in full swing, "time" has become a "timeless time" and "space" a "space of flows and connections." These are two features that identify what the sociologist Manuel Castells (2001) denominated "network society," a social structure for the twenty-first century proper; in other words, the society of information and knowledge. Nowadays, we therefore find ourselves before not only new leisure practices bound to virtual space, but also an authentic transformation as regards the whole leisure experience.

These variables of "timeless time" and "space of flows" have done nothing but transform leisure, and at this point in time we can see that there is no particular time set aside for the enjoyment of leisure, that is, a slot devoted to leisure outside working hours or the holiday period. As Dr. San Salvador del Valle (2009) suggests, leisure has reached the point where it is conceived as a true "fast leisure" which "involves an unbridled pursuit of taking full advantage of time perceived as a scarce commodity." In other words, taking into consideration the distinction made between "solid and liquid" society by the Polish sociologist and theorist of liquid modernity, Zygmunt Bauman (1999), finding itself enclosed in a "liquid society," leisure has adopted certain features based on constant change: acceleration, brief enjoyment of things, transience, frailness of human bonds, and constant uncertainty. This is a form of leisure based on a network society in perpetual movement where what really counts is being connected, in permanent connection with everybody else.

It is no wonder that most leisure practices are carried out through the Web, are social and interactive, and for instance, the use of virtual social networks has become one of the most widely practised digital leisure activities. These new leisure practices are located in the "simultaneous" and "timeless" virtual spaces which have come to form part of a culture of real virtuality and reality of the virtual. This is a connected society that, due to the unstoppable development of smartphones, has converged on an enjoyment of leisure detached from informal contexts and separated from working hours, adapting to the different and varied social times life allows us. Thus, just as the Argentine researcher Roberto Igarza states: "leisure has come to be regarded as small pleasure pills, brevities that can be enjoyed in the micro spaces left by labour activities or in the leisure fragments that the user allows him/herself during trips or in free time at home" (Igarza 2009, 43).

These are "leisure bubbles" which, for the author, form part of a leisure time that is interstitial, interactive, more selective, shorter lasting, and based on micro formats, and where entertainment, pleasure, relaxation, recreation, and fun are what matter. These are virtual spaces which at the same time become the ideal space for participation, collaboration, creation of collective knowledge, generation of projects, development of hobbies, and so on. They are new virtual leisure spaces which, far from establishing themselves as mere means of communication, socialization, entertainment, and fun, with prior education/training, can be spaces for learning and human development.

In summary, the features belonging to a liquid social context, based on postmaterialism values and influenced by the incessant advance of technology and development of a virtual space, have significantly varied conceptual understanding and the value placed on leisure. However, in turn, it has promoted new virtual forms of leisure which, grouped together here under the term "digital leisure," have become one of the most widely practised types of leisure by society as a whole, especially youngsters. That said, what do we refer to when we speak of digital leisure? What are these new forms of digital leisure like? What makes them different from traditional offline leisure?

DIGITAL LEISURE, DIGITALIZATION OF THE LEISURE EXPERIENCE

The use of ICTs and the Internet has changed the way in which people spend their leisure time. However, the development of Internet has not only provided us with new leisure activities but technology has also transformed people's individual access to leisure and has changed leisure behavior and experiences.

On the one hand, computers, smartphones, and tablets, among other technological devices, have become established as the screens par excellence in the leisure sphere, even prevailing over other leisure activities previously more widely practised; and on the other hand, cyberspace has turned into a virtual—but real—leisure space where a great deal of time is spent.

Both the various types of technological supports and virtual space have been established as important elements in the world of leisure, thus creating an emergent leisure bound to the use of technology and the Internet. This kind of leisure has been analyzed from various perspectives by a number of authors (Bryce 2001; Drotner 2008; Sánchez-Navarro, Tabernero, and Aranda 2009; García, López, and Sámper 2012; Nimrod 2012) and, in general, it is distinguished by "online leisure," "cyber leisure," "virtual leisure" or "e-leisure." These designations indirectly make a clear difference between spaces, distinguishing between offline leisure spaces (traditional leisure activities) and online leisure spaces (cyberspace). A distinction which therefore falls outside the scope of times and experiences by only distinguishing spaces.

For this reason, in this article we have chosen to use the term "digital leisure" in a broader sense as it is a concept that in some way we consider better encapsulates the transformation undergone by the sphere of leisure as a whole. That is to say, we consider that the digital age, apart from providing new forms of leisure linked to virtual spaces, has led to the digitalization of the leisure experience in global terms, not only as regards the different (physical or virtual) spaces, but also leisure contexts (tourism, culture, and recreation).

Therefore, we consider it important to make it clear what we are referring to when talking about "digital leisure," since focusing solely on the broadest sense of the term and linking it to all the opportunities for leisure that are subject to the possibilities brought to us by new technologies seems to be a rather ambiguous definition. One could understand as a leisure activity both "reading using an electronic book" and "participation in virtual communities" from so global a definition. These actions present clearly different characteristics depending on the space and device used. That is, in the former we are referring to a digitalized traditional leisure activity (reading), and in the latter a network leisure activity which, in essence, and due to its being carried out in a virtual space, acquires its own characteristics that make it different from other similar leisure activities practised up until now. Therefore, what do we understand by digital leisure and online leisure?

Let us base ourselves on the experiential philosophy of understanding the leisure concept explained at the beginning: we envisage "digital leisure" as being the digitalization of the leisure experience; a new dimension of leisure that we consider in keeping with a digital age which, as well as providing new forms of leisure, has also brought about an important change in living the leisure experience. So, we believe that in a society where typical values of liquid modernity prevail, the supremacy of hedonist values and the tendency towards technologization of life has, at the same time, penetrated the world of leisure (tourism, sport, culture, and recreation) as a whole. Life has become digital in numerous aspects, and leisure as a factor of human development has inevitably been unable to remain on the sidelines.

This is why, on the basis of the notion of being connected or unconnected and under the term "digital leisure," we include both traditional leisure activities that have been digitalized and new emergent forms of leisure linked to the virtual sphere (see Figure 10.1). By digitalized traditional leisure, we are referring to the traditional leisure activity, such as the game of Ludo/ Parcheesi, which can now be played online in this digital age; and as for the so-called new leisure forms, we consider the whole set of online leisure activities: virtual social networks, blogs, wikis, forums, virtual communities, and so on. In short, we understand digital leisure as a new paradigm of the leisure experience in which technology and the Web are present.

By means of this figure we intend to express that by "digital leisure" we understand the comprehensive digitalization of the leisure phenomenon, that is, leisure activities which, regardless of the time, space, and activity performed, have had their way of experiencing them modified. Therefore, when distinguishing between physical and virtual leisure spaces, we simply intend to refer to the creation of new leisure spaces and contexts that have made possible new opportunities for enjoying free time. That is to say: "We

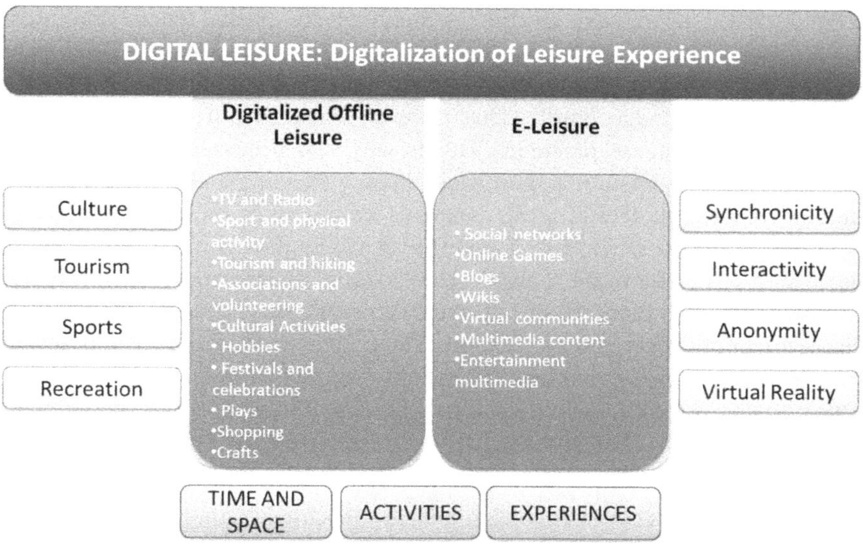

Figure 10.1

interpret digital leisure on the Internet not as being connected (online leisure) as opposed to unconnected (offline leisure), but as staying connected to the digitally constructed social world, not as virtual leisure versus real leisure, but as a new space for pursuing social relations" (García, López and Samper 2012, 397).

Nevertheless, we are aware that e-leisure activities developed in the virtual space in particular adopt their own characteristics that differentiate each one from another; features that Drs. Galit Nimrod and Hanna Adoni detail to perfection in the article "Conceptualizing E-leisure." According to these authors, Internet, apart from influencing traditional analysis variables for the sphere of leisure, such as "time and space," "activity," and "experience," has also helped add four new dimensions: "synchronicity," "interactivity," "anonymity," and "virtual reality." The latter four dimensions provide the Web leisure experience with a bonus.

We will start with an analysis of the three basic analytical aspects taken into consideration in the sphere of leisure until now: time and space, action, and experience. First of all, as analyzed in the previous section, leisure "time and space" has been undergoing considerable changes in recent times. Leisure today need not be necessarily practiced in out-of-work hours or at the weekend, and as a result, is ubiquitous and can be enjoyed at any time and in any place. Secondly, as we have been pointing out, leisure activities have also varied in essence. Moreover, due to the emergence of new leisure activities thanks to cyberspace, access to and enjoyment of these activities has also changed. Thus, activities carried out in the virtual space, for example, tend to be classified as "passive leisure," and conversely, offline leisure activities as "active leisure." Furthermore, online leisure, just like traditional leisure, has brought with it several barriers to participation; digital and social, usage, access, and knowledge hurdles that have an impact on freedom of choice when it comes to these practices. This is why it is necessary to include the following in this e-leisure: a valid basic literacy (reading and writing), media literacy (reading and writing in means of communication), literacy in the use of ICT and Internet, and a good level of English. Finally, as regards the leisure experience, the consolidated idea so intensively researched by Leisure Studies is maintained: that the degree of enjoyment depends on the degree of implication and motivation. Thus, apart from the time and space and the action to be carried out, the benefits derived from leisure remain the same (psychological, physiological, psycho-physiological, social, economic, and environmental). However, what contribution do the typically Internet-related features of "synchronicity," "interactivity," "anonymity," and "virtual reality" make to leisure?

One of the characteristics of cyber leisure is "synchronicity," defined by the psychoanalyst Carl Jung (1973) as the coincidence of two or more causally unrelated events whose meaningful content is identical or similar. We come across this feature in e-leisure activities very often, that is, it is possible to be watching television on both a television set and a mobile phone at the same time. However, what matters in the practice of online leisure activities is the degree of synchronicity that exists, that is, conversation by means of a chatroom could be synchronic or asynchronous. In other words, social networks have an important element of synchronicity but, it is possible that sending a message to a website, or sending a question at a certain time and receiving the reply days later, is asynchronous. In fact, it is precisely in the (high or low) degree of synchronicity that the quality of leisure experience in the virtual space lies, as it implies transformation into a symptom of power over time and space, which is highly valued by users.

The second trait exclusive to the new network-based forms of leisure is that of "interactivity," a term usually linked to ICT and one that adopts two major tendencies: one more strictly technological and another that is focused on the communicative value of interactivity. Francis Kretz makes the distinction between the software interactivity and human interactivity of interpersonal telecommunication services, possibly paving the way towards the coexistence of a more purely technological meaning of the term and another broader and somewhat more independent meaning of technology (Sádaba 2000). In any case, the strength of this characteristic attributed to the Web is that a person is involved, active and pursues a bidirectional transmitter-receiver conversation. The network by itself is not interactive, and it is therefore necessary for people to take action and provide feedback. Nimrod and Adoni state: "the more interactivity users experience, the more their use of the Internet experienced as leisure." Again participation and motivation are fundamental for benefiting from the interactivity that, by nature, characterizes the network space.

Thirdly, "anonymity" is another of the exclusive features that define the Web. From behind the screen people can define themselves as somebody with characteristics that are different from the reality, can create their identity and appear to be somebody that does not truly exist. And although this characteristic may be positive for certain people who wish to overcome personal or psychological problems, unfortunately, anonymity could lead to serious harm being caused through cybernetic harassment, cyberbullying, or identity theft. But finally, perhaps the most attractive feature offered by online leisure is the possibility of experiencing "virtual reality." A sensation of being present or existing in a place that is different from the place in which they "really" exist in their physical body at the same time (Rheingold 1991). Virtual space makes it possible to be in both spaces (offline and online) at the same time. On the one hand, it enables virtuality's integration into daily life, thus modifying the way we live leisure and, on the other hand, cyberspace allows the immersion of offline life in digital life. The virtual space acts just like another part of life yet fostering a continuation of leisure experiences, that is, making possible the expansion of offline leisure experiences to the virtual world and vice versa.

Stated briefly, basing ourselves on the philosophy of experiential leisure defended by leisure studies, we deem it appropriate to regard digital leisure comprehensively, not only as the type of leisure linked to the use of ICT and the Internet but also as a form of leisure geared to the twenty-first century. The influence of technology has been felt in its different spheres and has led to digitalization of the leisure experience. It is clear that in a network society with timeless time and the space of flows the ways of living leisure have changed, and that is why today it is more suitable to talk about fast, ubiquitous, and liquid leisure. Furthermore, virtual space has provided new online digital leisure activities that possess four characteristics making this type of leisure more appealing, synchronous, interactive, and anonymous, and which enable participation in a virtual reality.

REFLECTIONS TO RECAPITULATE

Without any doubt at all, the key idea that synthesizes this article is that "technology has reinvented twenty-first century leisure toward a digitalization of the experience." ICT and Internet have transformed the time, space, activities, and experiences of today's leisure and, in particular, the use of the social media, led by the virtual social networks of a social nature, such as Facebook, Twitter, and Tuenti (Spanish social network), has become established as one of the most widely practised and most positively rated by its users. Nevertheless, the central issue is that of verifying the quality of this online digital leisure and, on the basis of the experiential concept of leisure here presented, determining whether the use made of ICT and virtual space is dignified and of quality, or not. It will then be possible to decide whether the use made of these powerful tools of communication, socialization, and leisure tends towards leisure practices for mere entertainment and fun or enhances personal development.

However, data extracted at a worldwide level, as regards the penetration of ICT and Internet and Internet-use statistics, reflect a digital reality of a world in a state of continuous expansion. According to the most recent data (June 30, 2012) extracted from the "Éxito Exportador" website, which is based on data taken from the Nielsen Online and International Telecommunication Union (ITU) measurement companies, and World Stats, Internet use at world level is as high as 34.3%.[1] Asia is the leading continent as regards the number of Internet users (1,076.7 million), followed by Europe (518.5 million) and

North America (273.8). As for subscribers to Facebook, the leading social network in the world, Europe comes out on top with 243.2 million subscriptions, followed by Asia (236), Latin America (188.4), and North America (184.2). This data indicates a positive trend for the future, and that is what the chairman of Google, Eric Schmidt, recently published on Twitter: "everybody on the planet will be connected to Internet in seven years' time." But in the meantime, what use is really made of Internet, and is Internet a tool for entertainment or for development?

Taking as a reference the use that (Spanish) youngsters make of virtual social networks (VSNs), a tendency toward a superficial use can be seen. Although it is true that social media have enabled people/users to go from being simple receivers/consumers of digital content to actual producers, research indicates a reality that is totally divorced from the philosophy of a social and interactive Web. Recent studies carried out by various agencies (youth institutes, universities, and governments) reveal that the main aim of using VSN for the youth is that of communicating, socializing, and interacting; noting that entertainment and fun are the central axes in their online digital leisure practices. The lack of education on the useful uses and benefits of this media demonstrate that this is a spontaneous type of leisure based on the acquirement of immediate rewards and compensation, values that prevail in current society.

Nevertheless, apart from the uses attributed to the Web, we consider that what really counts is the personal and social benefit obtained from performing these new forms of leisure, and in this respect Internet has consolidated itself as the means of communication par excellence and the most highly valued by its users. Therefore, following the lines of Dr. Joe Bryce, researcher on this kind of leisure, we consider that this digital leisure is positive in a double sense because: "it provides disconnection and escape from the stress of everyday life . . . and can help improve our health and the construction of our personal identity" (Bryce 2001, 8).

After all, the key point is that these online new forms of leisure become established as positive new forms of leisure which people can enjoy and use to enhance their self-development; new forms of active, creative, and solidarity digital leisure that are references for identity, achievement, and justice, far from being mere entertainment and pastimes. This type of leisure, as well as being linked to hedonist values belonging to a consumer society in which brief enjoyment of things prevails, also offers opportunities for participation, collaboration, socialization, online learning, citizen empowerment, or generation of collective knowledge. To achieve this, however, prior education/training and literacy are necessary. Therefore, the challenge we launch is none other than that of placing our faith on fostering valid education in digital leisure which, on the one hand, trains a society to know how to enjoy their free time, and at the same time, ceases to regard Internet and VSN as mere tools for interaction based on banal conversation.

NOTE

1. This article is part of the doctoral thesis currently being carried out by the same author at the Leisure Studies Institute (University of Deusto), thanks to a pre-doctoral research fellowship grant awarded by the Basque government.

REFERENCES

Aranda, D., J. Sánchez-Navarro, and C. Tabernero. 2009. *Jóvenes y ocio digital. Informe sobre el uso de herramientas digitales por parte de adolescentes en España.* Universitat Oberta de Catalunya—UOC. Centre d´Estudis de l´Esplai (Fundació Catalana de l´Esplai). Transformemos el ocio digital. Un proyecto de socialización en el tiempo libre. Barcelona: Editorial UOC.

Bauman, Z. 1999. *Modernidad líquida*. Buenos Aires: Fondo de Cultura Económica.

Bryce, J. 2001. "The Technological Transformation of Leisure." *Social Science Computer Review* 19, 7–19.

Busquet, J., S. Ballano, D. Aranda, A. Medina, S. Morón, and L. Ruanao. 2012. E*l uso de las TIC y la brecha digital entre adultos y adolescentes. Primer esbozo del estado de la cuestión.* Grupo de investigación CONinCOM (Conflicto, Infancia y Comunicación) de la Facultat de Comunicació Blanquerna, Universitat Ramon Llull de Barcelona.

Castells, M. 2006. *La Sociedad Red*. Barcelona: Alianza Editorial.

Castells, M., I. Tubella, T. Sancho, and M. I. Roca. 2007. *La Transición a la Sociedad Red*. Barcelona: Ariel.

Cuenca, M. 2000. *Ocio Humanista. Dimensiones y manifestaciones actuales del ocio.* Documentos de Estudios de Ocio, 16. Bilbao: Universidad de Deusto.

Cuenca, M. 2004. *Pedagogía del ocio: modelos y propuestas*. Bilbao:Universidad de Deusto.

Cuenca, M. 2006. *Aproximación multidisciplinar a los Estudios de Ocio*. Documentos de Estudios de Ocio, 31. Bilbao: Universidad de Deusto.

Cuenca, M., A. Aguilar, and C. Ortega. 2010. *Ocio para innovar.* Documentos de Estudios de Ocio, 42. Bilbao: Universidad de Deusto. 2010.

Cuenca, J. 2008. *El ocio líquido. Un nuevo paradigma de experiencia.* En *La experiencia de ocio: una mirada científica desde los Estudios de Ocio,* Ed. Monteagudo Sánchez, 17–41. Documentos de Estudios de Ocio, 35 Bilbao: Universidad de Deusto.

Csikszentmihalyi, M. 1996. *Creativity: Flow and the Psychology of Discovery and Invention.* New York: Harper & Row.

Drotner, K. 2008. Leisure is hard work: digital practices and future competencies, youth, identity and digital media. *The MIT Press*. ?????

Ericschmidt. 2013. "For every person online, there are two who are not. By the end of the decade, everyone on Earth will be connected." #NewDigitalAge. Taken from: https://twitter.com/ericschmidt

García, E., J. López, and A. Samper. 2012. "Retos y tendencias del Ocio Digital: transformación de dimensiones, experiencias y modelos empresariales." En *ARBOR Ciencia, Pensamiento y Cultura* 188, 395–407.

Igarza, R. 2009. *Burbujas de ocio. Nuevas formas de consumo cultural*. La Crujia: Buenos Aires.

Jung, C. 1993. *Synchronicity: An Acausal Connecting Principle*. Princeton, N.J.: Princeton/Bollingen.

Monteaguado, M. J, ed. 2008. *La experiencia de ocio: una mirada científica desde los Estudios de Ocio*. Documentos de Estudios de Ocio, 35 Bilbao: Universidad de Deusto.

Nimrod, G., and H. Adoni. 2008. "Conceptualizing E-Leisure." *Loisir et Société / Society and Leisure*. 35(1), 31–56.

Kretz, F. 1983. "Dialogue, service, interactivité et leurs composantes: aspects de conception et d'évaluation." *Bulletin de l'Idate* 11, 77–103.

Observatorio Juventud de España y Centro de Investigaciones Sociológicas. 2011. *Cifras y jóvenes. Sondeo de opinión. Jóvenes y Tic*. Madrid: INJUVE.

Reig, D. 2012. *Socionomía: ¿vas a perderte la revolución social?* Barcelona: Deusto.

Rheingold, H. 1991. *Virtual Reality: Identity and community in cyberspace*. New York: Summit Books.

Sádaba, Mª R. 2002. "Interactividad y comunidades virtuales en el entorno de la world Wide web." *Comunicación y Sociedad*. Vol. XIII. Núm. 1, 139–166.

San Salvador del Valle, R. 2009. "El tiempo acelerado." *El País*.

Stebbins, R. 1982. "Serious Leisure: A conceptual statement." *Sociological Review*, 115–142.

Stebbins, R. 1997. "Casual Leisure: A Conceptual Statement," *Leisure Studies*, 16(1). DOI: 10.1080/026143697375485.

World Leisure Recreation Association. 1994. "*Carta para la Educación del Ocio*. En *El Derecho al ocio de las personas con discapacidad*," ed. S. Gorbeña; V. J. González. and Y. Lázaro, 243–252. Documentos de Estudios de Ocio, 4. Bilbao: Universidad de Deusto.

Chapter 11

Virtual Voyeurism and the Capitalization of Individuality on Facebook

Shahriar Kabir

The memory of the time is still not very obscure when the mass could be completely directed and driven, and the propagandas of the modern times can be taken as examples (for details, see Chomsky 1997, 7–57); however, at this age of postmodern reality, people tend to be more informed about the aspects that direct their lives; one such aspect is the social media, especially Facebook—the primary focus of this chapter. The invention of the social networking site Facebook.com is no more a recent phenomenon, but the vibe still seems to be at its puberty. Functional at a multitude of dimensions, this site has literally revolutionized the modes and methods of communication by transcending all the theoretical understandings of the very process of communication. People are massively occupied by it, but certain deviations can be observed from time to time. As a result, social media, especially Facebook, should be studied with a concern.

At this initial phase, the rationale behind this detailed discussion on Facebook requires clarification. It is not that Facebook is the only social networking site that needs to be studied with an exclusive focus. There are other social networking sites (popularly known as SNS) as well such as Twitter, Orkut, Google Plus, Instagram, Myspace, Friendster, and the like. Moreover, websites like these are not the only forms of social networking; in addition to these aforementioned sites, blogs and microblogs, content communities, collaborative projects, Internet forums, wikis, podcasts, video/photo sharing sites, review sites, virtual social worlds, and virtual game worlds also belong to the conceptual premises of social networking. Interestingly, no other social networking site has been as hugely successful as Facebook. As a result, a study of its backwashing effects would be more comprehensive.

To be more specific, this chapter concerns itself with a detailed understanding of virtual voyeurism and the capitalization of individuality on Facebook.

First of all, skeletal previews of the theoretical understandings of voyeurism and capitalization are provided in order to further the arguments. Voyeurism is a term exclusive to advanced-level feminism; the term capitalization is a part and parcel of Marxist understanding. This chapter hybridizes these two aspects in order to have a comprehension of the impact of the social networking sites, Facebook in particular. The next phase of this discussion separately looks at the reasons and consequences of these two negative aspects. Before drawing the conclusion, some recommendations would naturally emerge in the course of the discussion.

The range and scope of the term voyeur vary in relation to different genres as well as different disciplines. The concept of voyeur is an integrated part of the theoretical understanding of the genre pornography. In the course of the evolution of the genre, pornography—that was "once viewed as a medium simply devoted to sexual explicitness"—has now turned out to be about "gender relations and power: specifically patriarchal power and the continuing disempowerment and oppression of women" (Kipnis 1998, 153). Along with such notional change of the genre, with an addition to the social networking vibe, the term voyeur has also acquired newer dimensions (that are further elaborated in this chapter). Additionally, in behavioral science, the term voyeurism particularly refers to the erratic behavior of an individual. According to criminal psychology, it is synonymous to scoptophilia—"receiving sexual gratification by peeping through windows and so forth to watch people" (Hicky 2010, 143). A voyeur is commonly an individual with a psychological disorder. Fortunately, as Dr. Hicky explains, it is evident that most of the "'Peeping Toms' never progress past this [voyeuristic] deviant stage" (2010, 144).

As far as literary and cultural studies are concerned, two different approaches to voyeurism are available. First, in Freudian psychoanalysis, voyeur is associated with an active form of scopophilia. Secondly, and this reference needs elaboration, Laura Mulvey's (2004) seminal work "Visual Pleasure and Narrative Cinema" uses the concept of voyeur. She argues the audience of the narrative cinema is always presupposed as a male figure. In such a position, men become the bearer of looks, and women turn out to be the carrier of "to-be-looked-at-ness." Mulvey argues the image of women that evokes scopophilic pleasure can simultaneously evoke castration complex. In describing the evasion of this complex, she theorizes "two avenues to escape:" one is "sadistic voyeurism" meaning the "re-enactment of the original trauma" (investigating the woman, demystifying her mystery, making women guilty and then taking actions as punishment/forgiving), and the other is "fetishistic scopophilia" that connotes transforming women into fragmented part (e.g., the camera on breast), the cult or female star (2004, 837–48). At this point of her discussion, reflecting on the scoptophillics,

Mulvey defines voyeur as someone who "can only gain sexual satisfaction from watching an objectified other" (2004, 839).

The next concept capital—exclusive to Marxist understanding—is commonly understood as the economic investment that yields profit in a business venture, and capitalization denotes the process of turning something into capital; however, there is room for debate about the relevance of the theory itself, ergo the concept, as the practicality of Marxism is yet to be realized. The classic Marxist economic structure is comprised of a determining economic base and a number of determined cultural superstructures. In theory, Marx's dialectic sees the possibility of the formation of a classless society (Stumph 1975, 420–24). As Marx's prediction of a classless society is yet to be realized, the theory is often contested as impractical. However, in defense of Marxism, Lois Tyson (1999) answers the question of the relevance of the theory that "would *still* give us a meaningful way to understand history and current events" (emphasis added, 53). In addition, Pramod K. Nayar praises this theory as it "explores power relations embedded and concealed in cultural texts" (2011, 122). Peter Barry asserts "whereas other philosophies merely seek to understand the world, Marxism (as Marx famously said) seeks to change it;" the essential Marxist view is that "things are not 'innocent', but are 'determined'" (2002, 156–58). As a matter of fact, Marxism directly addresses the political power and the economic condition of a given scenario—this aspect makes the theory very lively and relevant for all the cases in relation to power struggle. As Marxism, especially the relation between capital and power perpetuation, provides the theoretical base, the discussion on the capitalization of individuality on Facebook is consistent with Marxist vocabulary.

The social networking site Facebook.com has many appearances. It is a form of cultural capital—"the possession of knowledge, accomplishments, formal and informal qualifications by which an individual may gain entry and secure a position in particular social circles, professions and organizations" (this definition is taken from Brooker 2003, 54). In addition to being an example of popular culture, the site has added to the definition of popular culture itself (three different definitions of popular culture are given in Storey 2002, 1–2). Although the concept of social networking has been in currency before Facebook, the site still marks the inception of a revolutionary step. Before the launching of the site, people were mostly concerned about chatting, and servers like yahoo.com and msn.com pioneered the concept of the social networking. Despite that fact, with the introduction of Facebook in September of 2005, the societal aspect of the virtual world has practically achieved a newer meaning.

Before knitting Facebook deeply with theoretical associations, it is necessary to mark its merits. The significance of the site can be argued on the basis of social, economic, and cultural dimensions; speaking point blank, this chapter focuses on the cultural aspect of Facebook. Besides, the site has practically revolutionized the concept of communication. The process of communication takes place mostly beyond the horizon of our conscious perception. Theoretically, it is fascinating to see a person disseminating the number of overt and covert messages while conversing with another person (further details can be found in Dimbleby and Burton 1996, 14–20). Not only did Facebook enhance our capability of communicating with each other, but also it has transgressed the categorical boundaries of the process. The experiences of communication can be divided into four categories: intrapersonal, interpersonal, group, and mass communication (Dimbleby and Burton 1996, 5–7). A careful observation shows that all these categorical divisions are dissolved on Facebook. An individual can engage in a monologue with oneself on the site; the mood of communication becomes dialogic when the very conversation accommodates another person. Group communication is an integrated option of the site whereas a message can be disseminated on a massive scale. In addition, features like status updating, sharing, tagging photos, chatting, ticker and so on are the aspects that facilitate such transgression of human limitation of communication.

The question is not what the changes on Facebook are; rather, how the changes are affecting the mass and/or would bring about devastating consequences demands attention. A simple status update, a tagged photo, or an event notification of communal value may seem innocent, but it may not be that simple. In addressing the relation between an individual in the virtual world, Lee Komito expresses his views likewise: "New technologies have opened a vast world of electronic communities to individuals, with the prospect of being able to experience cultural differences without travel or discomfort. *At the same time, the individual who enters such communities may remain a voyeur*" (emphasis added, 2011, 148). In fact, the process through which an activity of a user is perceived can posit the perceiver in the place of a voyeur. In most of the cases, people are interested in sharing personal affairs on Facebook that are intended to the near and dear ones. Unfortunately, one's overwhelming and inundating presence on the cyberspace constantly exposes one's personal affairs to complete strangers. The Facebook cover may be taken as an example. While updating their system with the cover photos, the site declared that the photo can be viewed by all existing users, whether they are on the friend list of a user or not. An individual may dislike the fact that an unknown person is able to see his/her personal photos. This is one of the reasons that the cover picture is either a group photo or an abstract one in many cases, especially when the user is a teenager. For such loopholes in their policies, Facebook is opening up the Pandora's Box and unleashing virtual voyeurism.

The next factor resulting from such virtual voyeurism is the ambivalence of identity. The identity of an individual rests in the dimensions of both the semiotic and the psychoanalytic. French Marxist Louis Althusser studies the interpellation of subject that ultimately leads to the filling up of a significant gap in Marxist thinking; he brought together psychoanalytic theories and semiotics in the Marxist analysis (Woodward 2002, 83). Althusser argues "all ideology hails or interpellates concrete individuals as concrete subjects" (2004, 699). Such a concrete form of identity is absent in cyberspace,[1] ergo in cyberculture.[2] At the backdrop of a virtual world, where the somatic body could be rendered nonessential, the formation of identity suffers from the absence of some essential formative elements that constitute our perception of the matrix of identity. Thus, in such a posthuman form, the individual capabilities are practically extended to a certain extent. Unfortunately, these capabilities have negative effects like fake identities, online theft, and, of course, voyeurism as far as Facebook is concerned.

The ambivalence of identity is not only facilitating voyeurism, rather, the disturbing fact is that it is contributing in the formation of a generation to whom the social and cultural prohibition of voyeurism would not carry much gravity; as a result, the trend would instill and integrate sociopathic tendencies into the posterity. Martha Stout defines a sociopath as an individual with no feeling of guilt or remorse, no limited sense of concern for the well-being of strangers, friends, or even family members (2005, 1). The social networking facilities are aiding at the promotion of the sociopath mentality, as the norms and formulations that shape our psychology are being loosened. This scenario is thus creating rooms for unforeseen anomalies. More haunting is the fact that Facebook or any other social networking sites would never require the physical presence of a body in order to be identified as an individual—this particular feature of the cyberspace is a marked paradigm shift in an age of identity crisis. With the freedom from the bodily barriers, a particular group of users would naturally be drawn to voyeuristic activities as the nature of prying is one of the most primitive of all forces. The resurgence of such a nature might not cause upheavals immediately, but the probable consequences would be grave.

With virtual voyeurism looming large, Facebook is enabled to serve the purpose of social surveillance or electronic panopticon. The function of panopticon, as Foucault has put it, is "to induce in the inmate a state of conscious and permanent visibility that assures the automatic functioning of power . . . that the perfection of power should tend to render its actual exercise unnecessary" (Huggard 2002, 196). The French thinker very craftily outlined the ambivalence of panopticon—the visual of a panopticon evokes servile nature in prisoners; there might not be any guard in a panopticon, but the structure itself can be deemed as an extension of the physical presence of authority.

Akin to that, Facebook has got such an aspect integrated into its system very recently—the ticker feature. At times, an individual might want to surf around on Facebook with no intention of being disturbed. The ticker feature just acts as an agent that would declare the online presence of that person. One personal observation is that people, at the oddest of times, would knock just to ask why I am awake as I liked a photo or a status that showed on his/her ticker. I may not be the only person to have such an experience, and I strongly believe a lion's share of Facebook users would not appreciate such an unnecessary courtesy. Therefore, Facebook nowaday feels like a prison, and the ticker feature couldn't have been more perfect an entrapment.

The scope of virtual voyeurism is creating rooms for the users to be voyeurs by instilling less cultural prohibition on voyeurism, and schizophrenic mentality would flourish if preventive measures are not taken immediately; the consequences could be unforeseen and definitely devastating. Based on real-life observations and studies, Beverly Engel categorizes seven types of "Jekyll and Hyde" (2007, 15–16); however, the fluctuating matrix of identity on cyberspace makes it hard to determine the forms of schizophrenia virtual voyeurism would produce. One common symptom is that a user is may be completely different in social networking sites from what s/he is in real life. An individual may be very outspoken on the cyberspace, whereas that very person could be practically timid and shy during face-to-face interaction with people—this would be one form of schizophrenia.

Another form of schizophrenia is gender confusion. Many existing Facebook users represent their profile to be of the opposite gender, commonly known as fake profile. It must be noted that a fake ID and a fake profile is not the same. A fake ID means an account that replicates a real person that is very consistent with the original one. On the other hand, a fake profile is created with an amount of false information. Representing oneself either as a female or a male figure is common in fake profiles. Characteristically entertaining, these forms of representation still mirror schizophrenic tendency, and I am afraid that this would not be the only one. In a nutshell, for a generation heavily (and, at times, completely) dependent on social networking, the line between sanity and personality disorder would be thinned to a great extent. The social and cultural consequences would therefore be undoubtedly devastating.

The comprehension of the capitalization of individuality would require the understanding the status of an individual in virtual communities. On Facebook, an individual becomes the unconscious victim of the process of commoditization. More disturbingly, the process only conceals the "money-form of the world of commodities," and it never discloses "the social character of private labour, and the social relations between the individual producers" (Marx and Engels 2001, 779). Clearly, the individual user becomes an active agent and furthermore loses his/her significance as a singular entity in a virtual community.

Interestingly, one may claim that one is able to control the temptation of providing personal information on Facebook; however, if Lacanian proposition is taken into account, the process takes an interesting turn. Lacan proposes that "'the phallus represents a moment of division [that 'lack-in-being] which re-enacts the fundamental splitting of the subject itself'" during the imaginary phase of the psychic construction of an individual (Homer 2007, 57). His theory further explicates that the very split within the subject facilitates the functioning of numerous activities that preach wholeness. This very pattern of psychological exploitation is seen on Facebook. After an individual opens an account, s/he is required to provide his/her details. One may leave one's profile incomplete, but the site consistently asks to update the profile. Providing details of schools attended, movies watched, TV series enjoyed, music listened to may be fun at the beginning. However, after the initial ones, the questions become quite intrusive like marital status and unnecessary (for me at least) details of family members. These sites show a very *irritating* profile completion percentage that feeds on the Lacanian notion of completeness or wholeness. Facebook users tend to complete their profiles, and thus their details are ready to be used as a corporate capital. Moreover, speaking point blank, a completed profile always allures voyeurs—a disturbing fact that currently has no remedy.

In order to explicate the process of the capitalization of individuality, the promotional aspects of the website need to be taken into consideration. The first and foremost example would be the usage of numerous applications based on Java platform. An application (popularly known as app) may be a game or any calculative aspect like the percentage of love, luck, profile visits per day and so on. Whatever the application may be, they all use similar terms and conditions, and a user is required to accept those. The most common demand of those applications is the permission to access basic (read: personal) information. After that, these applications require an optional permission to post on the user's behalf on their respective walls. Things can be managed up to this extent, but right after that, comes the most peculiar twist. These applications require participation of other users in order to progress and unlock further achievements. Usually, these applications are addictive, and the user frequently asks his/her friends to subscribe to and use that application.

Subscribing to or using an application is not a matter of concern, rather the important point is these applications on Facebook access the personal information. Once an application is used, there are more and more suggestions for similar types of applications. Sometimes the demand of such applications goes viral. One might remember the "Farmville series" that once practically occupied most of the notifications. This particular application had demanded not only the participation of other mutual friends, but also it literally compelled the users to start sharing things with complete strangers on Facebook. Apart from the applications that go viral, there are other applications that constantly search for personal information like birthdays, anniversaries and so on. Most of these suggestions of these applications are automatically generated for a user, but the disturbing fact is that the access to an individual's personal information and frequently visited pages is being utilized in order to provide statistical accuracy. This is not expected; however automated or manually generated, the raw data are still somebody's personal information. These social networking sites use personal information, and they are investing that information as a capital for corporate gain.

The final instance of the capitalization of individuality can be understood through the system of logging in to other third-party websites through a Facebook account—commonly known as social login. Websites like 4shared.com, Imdb.com, Academia.edu and so on offer such login systems, and this process is getting popularity for downloading contents from the internet archives or just signing in to other websites. In this process, one single account is getting exposed to numerous websites, and exclusivity of personal information is at stake as a result. According to Gigya.com, 51 percent of people prefer social login via their Facebook account. The percentage is 66 in the case of mobile devices, 74 percent in e-commerce, 61 percent in consumer brands, 50 percent in travel/hospitality, and 81 percent in case of educational or non-profit purposes (White 2013). With such a viral rate, the personal detail is undoubtedly at the risk of being exploited. In practice, the details are used in order to advertise customized products that reduce the necessity of campaigning for the product itself—something that should not be allowed for our own sakes.

Achieving a conclusive stance on the capitalization of individuality would be partially incomplete without Fredric Jameson. In his ground-breaking essay "Postmodern and Consumer Society," Jameson views postmodernism as "a periodizing concept whose function is to correlate the emergence of new formal features in culture with the emergence of a new type of social life and a new economic order" (2001, 1962). By the end of the essay, he asserts that the "informational function of the media would thus be to help us forget, to serve as the very agents and mechanisms for our historical amnesia" (Jameson 2001, 1974). In the case of Facebook, this proposition can be seen in action. The exhaustive personal details seem to be incomplete forever. Facebook has changed its privacy policy, appearance, and mode of execution so much that one finds it easier to accept those. Accommodation of such changes comes with very little resistance, and no one seems much bothered about it anymore—a very discreet sign of collective amnesia that Jameson has once explained.

Another unfortunate fact in the case of capitalization of individuality lies in its uncontested hegemony. Antonio Gramsci describes hegemony to be the "'spontaneous' consent" given to the "dominant fundamental group" determined by their "position and function in the world of production" (2011, 12). It is the discursive hegemony that is resulting the capitalization of individuality. An individual and his/her private information should be valued. Vincent Campbell informs about an amendment of "protecting individual citizens as much as large news media organisations" in journalism; he further adds that the "market-driven news media" fails to do so in terms of "choice and access, diversity and representation, and paternalism versus populism" (2004, 55–56). Moreover, a "deontological principle that emphasizes personal privacy for the sake of human dignity can serve as a foundation for our thinking and action. *This perspective is anchored in the person whose privacy is at stake, rather than beginning and ending with the judgment of those who are using the personal data on their own behalf*" (emphasis added, Christian et al. 2006, 130). The respective quotes are given here as there should at least be some kind of amendments and/or guiding principles for cyberspace (especially in case of social networking) as well. An individual must not be deemed as a consumer. If s/he wishes to be so, there must be some specific sets of standards and procedures, so that one can be aware of the further consequences (Yes! Akin to the software license agreements that no one, I earnestly believe, has ever read). What we currently have is a farcical amount of literature in the name of policy. If this is allowed to continue, there would be a *no-turning-back* point. Individual information would be used for corporate gain—a scenario that no one would (or should) appreciate at all.

While drawing the conclusion, it must be restated that the prime focus of this chapter is the privacy issues specifically related to social networking sites with an exclusive focus on Facebook. Facebook has been perused to make the discussion comprehensive as its users are everywhere. Undoubtedly, the site has brought about many revolutionary changes, both individual and collective. Unfortunately, behind the all-consuming presence of goodness, some issues have been overlooked so far; this chapter has (un) covered two such issues: virtual voyeurism and the capitalization of individuality. As this discussion explains, Facebook's almost aggressive execution policy is intruding into personal lives, and the loopholes in the system are facilitating some systematic corruptions. The thin line between *allowed* and *prohibited* is dissolving. If this continues, then the consequence, as this chapter elaborates, would be extremely serious. It is virtual voyeurism that should be prevented at first while the preventive measure (might) still be effective. If that happens, then the capitalization of individuality would be easier to prevent.

NOTES

1. The concept of cyberspace has a "complex history of technological innovations, including cybernetics, virtual reality and sociological and epistemological reactions to new forms of physical and urban space, as well as being a familiar staple of science fiction throughout the century" (Whittaker 2003, 135).

2. A combination of the modern technological advancement, a virtual world, and the world of hypertext and hyperspace altogether creates a unique cyberculture that denotes "the vast gathering of information, misinformation, sounds, images and ideas that can be accessed through the internet, along with the set of practices, attitudes, values and ways of thinking" (Edgar and Sedgwick 2008, 84–87).

REFERENCES

Althusser, Louis. 2004. "Ideology and Ideological State Apparatus." In *Literary Theory: An Anthology*, 2nd ed. Edited by Julie Rivkin and Michael Ryan, 693–702. Oxford, UK: Blackwell Publishing.

Barry, Peter. 2002. *Beginning Theory: An Introduction to Literary and Cultural Theory*, 2nd ed. New York: Manchester University Press.

Brooker, Peter. 2003. "Cultural Industry." In *A Glossary of Cultural Theory*, 2nd ed. New York: OUP.

Campbell, Vincent. 2004. *Information Age Journalism: Journalism in an International Context.* London: Arnold.

Chomsky, Noam. 1997. *Media Control: The Spectacular Achievements of Propaganda.* New York: Seven Stories Press.

Christians, Clifford. G., Kim B Rotzoll, Mark Fackler, Kathy Brittain McKee, and Robert H. Woods. 2006. *Media Ethics: Cases and Moral Reasoning*, 7th ed. New Delhi: Pearson Education.

Dimbleby, Richard, and Graeme Burton. 1996. *More than Words: An Introduction to Communication.* New York: Routledge.

Edgar, Andrew, and Peter Sedgwick, eds. 2008. "Cyberculture." In *Cultural Theory: The Key Concepts*. Oxfordshire: Routledge.

Engel, Beverley. 2007. *The Jekyll and Hyde Syndrome.* New Jersey: John Wiley & Sons.

Gramsci, Antonio. 2011. *Selections from the Prison Notebooks of Antonio Gramsci.* Edited and Translated by Quintin Hoare and Geoffrey Nowell Smith. New Delhi: Orient Blackswan Private Limited.

Haugaard, Mark. 2002. *Power: A Reader.* UK: Manchester University Press.

Hickey, Eric. W. 2010. *Serial Murderers and Their Victims*, 5th ed. USA: Wadsworth, Cengage Learning.

Homer, Sean. 2007. *Jacques Lacan.* Oxon: Routledge.

Jameson, Fredric. 2001. "Postmodern and Consumer Society." In *The Norton Anthology of Theory and Criticism*, edited by Vincent B. Leitch and Jeffrey J. Williams, 1932–1975. New York, NY: W. W. Norton & Company.

Kipnis, Laura. 1998. "Pornography." In *The Oxford Guide to Film Studies*, edited by John Hill, and Pamela Church Gibson, 153–157. New York: OUP.

Komito, Lee. 2011. "The Experience of Virtual Communities: Cosmopolitan or Voyeur?," In *The Cause of Cosmopolitanism: Depositions, Models, Transformations*, edited by Patrick O'Donovan, and Laura Rascaroli, 135–150. Switzerland: Peter Lang.

Marx, Karl, and Freidrich Engels. 2001. "The Fetishism of Commodities and the Secret Thereof." In *The Norton Anthology of Theory and Criticism*, edited by Vincent B. Leitch and Jeffrey J. Williams, 776–783. New York, NY: W. W. Norton & Company.

Mulvey, Laura. 2004. "Visual Pleasure and Narrative Cinema." In *Film Theory and Criticism: Introductory Readings*, 6th ed. Edited by Leo Braudy and Marshall Cohen, 837–848. New York, NY: OUP.

Nayar, Pramod K. 2011. *Contemporary Literary and Cultural Theory: From Structuralism to Ecocriticism.* New Delhi: Dorling Kindersley (India) Pvt. Ltd.

Storey, John. 2002. *Cultural Theory and Popular Culture: An Introduction*, 3rd ed. UK: Prentice Hall.

Stout, Martha. 2005. *The Sociopath Next Door.* New York, NY: Broadway Books.

Stumpf, Samuel Enoch. 1975. *Socrates to Sartre: A History of Philosophy*, 2nd ed. USA: McGraw-Hill Book Company.

Tyson, Lois. 2006. *Critical Theory Today: A User-Friendly Guide*, 2nd ed. New York: Routledge.

White, Victor. 2013. "The Landscape of Social Login: The Identity War Heats Up." *Gigya.* http://blog.gigya.com/social-login-infographic-q3-2013/.

Whittaker, Jason. 2003. *The Internet: The Basics.* New York: Routledge.

Woodward, Kath. 2002. *Understanding Identity.* London: Arnold.

Chapter 12

Are Google's Executives Liable for Uploaded Videos?

Italian Case

Anna Rita Popoli

YOUTUBE

History

Social network sites are web-based services that allow individuals to construct a public or semipublic profile within a bounded system, articulate a list of other users with whom they share a connection, and view and traverse their list of connections and those made by others within the system. The nature and nomenclature of these connections may vary from site to site (Boyd and Ellison 2007, 211).

In recent years, the Internet has experienced enormous increases in traffic from social networking media data and video streaming on-demand web-based services because of the increasing interest in user-generated content (Ameigeiras 2013, 360). Among all audio and video sites, YouTube has became the most dominant, being rated as the third most visited Internet site (Alexa Corporation 2013).

YouTube.com is a video sharing site, which offers embedded social networking features. In much the same manner as other media sharing sites (e.g., Flickr, Picasa, Treemo), it allows users to calibrate their participation and interaction levels—alternating between viewing the site's content passively and actively communicating with other users (Rotman and Preece 2010, 318–20).

YouTube was registered on February 15, 2005, by Steve Chen, Chad Hurley, and Jawed Karim, three former employees of the American e-commerce company PayPal. They had the idea that ordinary people would enjoy sharing their "home videos" (Wikipedia 2013).

The site was officially launched in December 2005 and immediately attracted a huge number of users. The founders made a deliberate decision not to include advertising as part of the site. Funds, however, came from other sources. In November 2005, the company received almost 10 million dollars in funding from Sequoia Capital, which had previously helped finance Apple, Google, and other Silicon Valley companies (Wasko and Erickson 2009, 373–74).

In 2005 the American search engine company Google Inc. had launched a video service, Google Video, but it failed to generate much traffic, and Google was prompted to purchase YouTube in November 2006. Rather than merging the web sites, howevexr, Google continued YouTube's operation as before. Thus, from the beginning, Google's intention seemed clear: to develop YouTube's potential for attracting advertising revenues. Moreover, to reduce the risk of copyright-infringement lawsuits, Google negotiated deals with a number of entertainment companies that would allow copyrighted video files from YouTube (Hosch 2013).

Since YouTube is a subsidiary of Google Inc., it remains important to examine the parent corporation more carefully. Founded in 1998 by Larry Page and Sergey Brin, Google began as a "Web crawler" or a search engine that traverses the Web in search of requested information. The company grew rapidly. Google's organization currently (2009) includes various divisions, such as Google.com—Search and Personalization; Communication, Collaboration, and Communities; Downloadable Applications; Google GEO—Maps, Earth, and Local; Google Checkout (online shopping); ànd Google Mobile. The main google.com site has been expanded to include special features, such as Image Search and Book Search; Google Scholar provides a simple way to do a broad search for relevant scholarly literature, including peer-reviewed papers, theses, books, abstracts, and articles. Google's goal is to organize the world's information and make it universally accessible and useful. To do this, the company relies on advertising to generate revenues. Their advertising strategies include content-targeted ads on google.com, ass well as programs such as AdWords and AdSense, which help content owners to monetize their content by adding advertising to content, as well as other advertising strategies. Google is a public corporation—meaning their stock is available to the public. At the end of 2007, the company's two founders and the CEO (Eric Schmidt) owned almost 90 percent of outstanding Class B common stock, representing more than two-thirds of the voting power (Wasko and Erickson 2009, 374–75).

Although media corporations CBS, the BBC, Vevo, Hulu, and other organizations offer some of their material via the site, as part of the YouTube partnership program, most of the content on YouTube has been uploaded by individuals.

In 2007, YouTube began trials of a system for automatic detection of uploaded videos that infringe copyright. The system, which became known as Content ID, creates an ID file for copyrighted audio and video material, and stores it in a database. When a video is uploaded, it is checked against the database, and the video is flagged as a copyright violation if a match is found. When this occurs, the content owner has the choice of blocking the video to make it unviewable, tracking the viewing statistics of the video, or adding advertisements to the video.

YouTube has also faced criticism over the offensive content in some of its videos: for example for defamation, pornography, and material encouraging criminal conduct (Wikipedia 2013).

Structure

YouTube allows registered users to upload their videos and share them with the public: to view a video, one does not need to register or identify in any way, however, in order to broadcast a video, a user needs to create an account. Registered users can choose between publishing their videos publicly, or privately so that only invited viewers can watch them. When posting a video, YouTube users are asked to briefly describe and tag it. Most descriptions are short and consist of a paragraph or less. Tags are used to relate information about the video's topic and contents. Tags are left to the user's discretion, and are not subject a predefined taxonomy. YouTube's search engine retrieves videos according to the search terms found in the video's description, tags, or user name (Rotman and Preece 2010, 320–22). Registered users can upload videos up to 15 minutes each in duration. Users who have a good track record of complying with the site's community guidelines may be offered the ability to upload videos up to 12 hours in length, which requires verifying the account, normally through a mobile phone. YouTube accepts videos uploaded in most container formats (Wikipedia 2013).

Right from the start, the YouTube interface offers various choices. In addition to the search facility it shows which videos are being played at that very moment, it presents a number of "promoted videos" (proposed by the YouTube company) as well as "featured videos" that are highlighted for their qualities.

In order to search the database for video clips, the YouTube interface, to begin with, includes a common search bar. However, the Application Programming Interface (API) also offers the possibility to automate search processes and retrieve videos according to a certain search string in order to implement them into a different application. As mentioned, YouTube's information management relies on machine-readable information describing the video clips that permits their retrieval according to key terms. This so-called meta-information is initially generated by the users uploading the videos and consists of the title users give a clip, the information added to an info box in order to provide back-ground information or a summary of the clip, and the tags, that is a number or keywords one can select freely according to what one assumes to be appropriate labels for these images. Users viewing videos also provide meta-information implicitly, since the viewing rate is a criterion for the order of videos that match a given keyword. The activities of users, either those supplying information about videos they upload in the form of a title, additional information, and tags, or those viewing, rating, commenting on or flagging videos, do affect the responses of YouTube to search requests.

All these operations that YouTube offers its users—or rather which must be used for YouTube to generate the metadata necessary for its functioning—are at first sight ancillary options and additional services. Quite on the contrary, however, they actually provide the indispensable basis for the database's information management.

Finally, YouTube also offers the possibility for a very specific type of user comment, namely the so-called flag button for "reporting the video as inappropriate." This option was inspired by the idea that this might help the efficient and rapid removal of pornographic, racist, or otherwise extremist content from the site.

Terms of Service

At the bottom of the every single Web page of YouTube there are "Terms," "Privacy," "Policy and Safety," "Send feedback" and "Try something new!." In this paper the analysis is limited to "Terms" and "Privacy." Terms of service states in article 1.1 that the use of the YouTube website is subject to the terms of a legal agreement between the user and YouTube: YouTube means YouTube LLC whose principle place of business is in San Bruno, California, United States.

By using the Service the Terms are accepted (2.2), but user may not accept the Terms if (a) he/she is not of legal age to form a contract, or he/she is a person who is either barred or otherwise legally prohibited from receiving or using the Service under the laws of the country in which he/she is resident or from which he/she accesses or uses the Service (2.3).

Article 4.3 states that the user agree that he/she will be solely responsible (to YouTube, and to others) for all activity that occurs under his/her YouTube account.

Article 5 concerns general restrictions on use. Of particular interest for this research are article 5.1.9 and article 5.6. The first states "You agree not to collect or harvest any personal data of any user of the Website or any Service (and agree that this shall be deemed to include YouTube account names); while the second "You agree that you are solely responsible for (and that YouTube has no responsibility to you or to any third party for) any breach of your obligations under the Terms and for the consequences (including any loss or damage which YouTube may suffer) of any such breach.

The following article 6 concerns copyright policy.

An in-depth analysis of article 7 ("Content") is necessary for this paper. Each account holder may submit content. However, YouTube does not guarantee any confidentiality with respect to content (7.1). User retains all his/her ownership rights in his/her content, but he/she is required to grant limited license rights to YouTube and other users of the service (7.2). User understands and agrees that he/she is solely responsible for his/her own content and the consequences of posting or publishing it. YouTube does not endorse any content or any opinion, recommendation, or advice expressed therein, and YouTube expressly disclaims any and all liability in connection with content (7.3). Article 7.4 states that user represents and warrants that he/she has (and will continue to have during his/her use of the service) all necessary licenses, rights, consents, and permissions. Moreover user agrees that he/she will not post or upload any content which contains material which it is unlawful for the country in which he/she is resident (7.6). Then content will not contain any third-party copyright material, or material that is subject to other third-party proprietary rights (including rights of privacy or rights or publicity) unless the user has a formal license or permission from the rightful owner, or is otherwise legally entitled, to post the material (7.7). Finally user understands and acknowledges that in using the service, he/she may be exposed to content that is factually inaccurate, offensive, indecent, or otherwise objectionable; the user agrees to waive, and hereby does waive, any legal or equitable rights or remedies he/she has or may have against YouTube (7.9).

Other articles concern: "Rights you licence" (8); "YouTube content on the Website" (9); "Links from YouTube" (10); "Ending your relationship with YouTube" (11); "Exclusion of Warranties" (12); "Limitation of liability (13); and "General legal terms" (14) (YouTube 2013).

Privacy Policy

By clicking on "privacy" in a YouTube' page the user literally jumps in Google's privacy policy. It is interesting to note that this happens with all of Google's products: however, among these YouTube is the only with is own terms and conditions. Chrome, Books, Wallet, and Fiber have additional information concerning privacy.

The privacy policy explains: what information Google collects and why the collection is done; how Google collects that information; and the offered choices, including how to access and update information.

First, the privacy policy informs that data are collected in two ways: information given directly by the user, and information obtained from the use of

Google's services. The last includes: device information—such as hardware model, operating system version, mobile network information, log information—for example details of how Google's services are used, phone number, time and date of calls, duration of calls; Internet protocol address—such as browser type, browser language, system activity, cookies that may uniquely identify user's browser, location information, unique application numbers, local storage, cookies, and anonymous identifiers.

The reason of the collection is explained in few words: "we collect information to provide better services to all of our users."

Secondly, collected data are used to provide, maintain, protect, and improve Google's services; acquired information also is used to offer tailored content to Google's users. Moreover, Google reserves the right to combine personal information from one service with information, including personal, information, from other Google services. Furthermore, Google's privacy policy informs that: "Google processes personal information on our servers in many countries around the world. We may process your personal information on a server located outside the country where you live."

Thirdly, privacy policy explains which are users' rights: review and control, view and edit, control of shared information (Google 2013).[13]

However, in any part of the privacy policy content providers are advised that they had to request permission from third parties to post material that is protected by the privacy directives. This is a fundamental point, as the basis of the decision of the Tribunal of Milan is the lack of this information to Google's users.

ITALIAN LEGISLATION CONCERNING PRIVACY AND ELECTRONIC COMMERCE

The Web Evolution

In the context of the Web 2.0, the relationship between providers of Web hosting and addresses of such a service has significantly changed. Originally, Web hosting consisted mainly in websites (html pages and related documents) completely developed by the recipient of the hosting service, including the way the content was posted, the structure of the websites and so on. The host provider only made available the server (disk-space and processor) for storing the website, the connection from that server to the Internet, and the software (the Web-server) that would provide access to the website (by typing a domain name or using a search engine). While the recipient of the service had the greatest freedom in developing the website according to his or her tastes and preferences, editing Web pages was relatively difficult and complicated, and thus the Web could not be a creative space for the majority of people (Sartor and Cunha 2010, 374).

Web hosting has dramatically changed in the last years. Now platforms are available that facilitate the creation and distribution of online contents, thus enabling everyone to participate in these activities. Among the most popular platforms used worldwide we can name iTunes and YouTube for videos, Facebook for personal information, Wordpress for blogs, Twitter for short messages, e-Bay for auctions; the list is far from being exhaustive. Such platforms to a different degree support the creative activity of the users: first, they provide facilities (and constraints) for creating content, such as page templates, ways to organize the information and link it, apps for an infinite variety of functions; secondly, they facilitate the retrieval of the user-generated materials, by indexing, classifying, and ranking them (usually by aggregating users' preferences and choice). Such platforms are mostly run by commercial companies that usually make a profit by associating advertisements to the user-generated materials, often by selecting the ads on the basis of the content of such materials (Cunha, Marin, and Sartor 2012, 51).

Google, whose mission is "to organize the world's information and make it universally accessible and useful" (Google 2013), exemplifies this business model.

The Web 2.0 represents an evolution whose scope goes beyond computing and the Internet, to reach the whole system of media, information, and communication.

Unfortunately, the production and distribution of user-generated content is not always socially beneficial: consider for instance, defamation, violation of intellectual property, support of criminal activity, incitement to hate, child pornography, etc. (Cunha, Marin and Sartor 2012, 51).

Data Protection Italian Legislation

Italian legislation concerning data protection and electronic commerce—both from European directives—was conceptualized and implemented before the advent of the user-generated web.

The code on personal data protection, referred to for brevity as privacy code, is divided into three parts: the first includes the general provisions applicable to all types of processing, integrated by further rules for public subjects and others for private and public bodies; the second displays the provisions applicable to specific sectors; finally, the third part includes the regulations on the forms of protection, the sanctions, and the office of *Garante per la protezione dei dati personali* (Italian Data Protection Authority). The code stands as the transposition not only of most of directives 95/45/

EC[1] and 2002/58/EC,[2] but also of the decisions and opinions issued by the Italian Data Protection Authority.

The starting point is the notion of "personal data," that is "*any information relating to natural or legal persons, bodies or associations, that are or can be identified, even indirectly, by reference to any other information including a personal identification number*" (article 4 (1), letter b).

From the subjective point of view, a part data subject, privacy code defines the data controller as "*any natural or legal person, public administration, body, association or other entity that is competent, also jointly with another data controller, to determine purposes and methods of the processing of personal data and the relevant means, including security matters*" (article 4(1), letter f); as data processor "*anyone that processes personal data on the controller's behalf*" (article 4(1), letter g); finally, as person in charge of the processing "*the natural personal that have been authorized by the data controller or processor to carry out processing operations*" (article 4(1), letter d).

Within the definition of personal data it is important to distinguish the notion of "sensitive personal data," that is, "*personal data allowing the disclosure of racial or ethnic origin, religious, philosophical or other beliefs, political opinions, membership of parties, trade unions, associations, or organizations, of a religious, philosophical, political or trade-unionist character, as well as personal data disclosing health and sex life* (article 4(1), letter d).

Finally, as a general rule, personal data processing consists of "*any operation, or set of operations, carried out with or without the help of electronic or automated means, concerning the collection, recording, organization, keeping, interrogation, elaboration, modification, selection, retrieval, comparison, utilization, interconnection, blocking, communication, dissemination, erasure and destruction of data, whether the latter are contained or not in a database*" (article 4(1), letter a). To simplify, any operation performed on personal data represents "processing" under the Italian regulation.

The legislator outlined a strict regime of third-party liability. Indeed, "*whoever causes damage to another as a consequence of the processing of personal data shall be liable to pay damages pursuant to article 2050 of the Civil Code*" (article 15(1). The set of sanctions is also integrated by criminal offences consisting of "*Unlawful data processing*" (article 167), "*Untrue declarations and notifications, submitted to the Data Protection Authority*" (article 168), "*Failure to comply with the safety measures*" (article 169), and "*Failure to comply with provisions issued by the Data Protection Authority*" (article 170). The set of sanctions was completed by the presence of several administrative sanctions: "*Providing no or inadequate information to data subjects*" (article 161); "*Other types of non compliance*" (article 162);

"*Failure to submit notification or submitting incomplete notification*" (article 163); "*Failure to provide information or produce documents to the Data Protection Authority*" (article 164) (Popoli 2012, 8–10).

E-Commerce Directive and Italian Legislation

With regard to liabilities for the illegal processing of personal data, the provisions of the legislation concerning privacy would seem that they have to be coordinated with those of the e-Commerce Directive.[3] The European Union e-Commerce Directive adopts the definition of

> Information Society Service under Article 1.2 of Directive 98/34/EC to refer to ISPs and addresses the civil and criminal liabilities of ISPs acting as intermediaries. The Directive provides that ISPs will not be held liable in any field of law in which an application of strict liability would impair the expansion of electronic commerce within the E.U. The limitation on liability provided for by the Directive are established in a horizontal manner, meaning that they cover liability, both civil and criminal, for all types of illegal activities initiated by third parties.[4]

Article 2 of the e-Commerce Directive considers a service provider as any natural or legal person delivering "*any service normally provided for remuneration, at a distance, by electronic means and at the individual request of a recipient of services*" and a recipient of service as "*any natural or legal person who, for professional ends or otherwise, uses an information society service, in particular for the purposes of seeking information or making it accessible.*"

In this case, Italian legislation has its basis on European regulation: Legislative Decree n. 70 of April 9, 2003 implemented the EU Directive on Electronic Commerce. Article 14 of the Legislative Decree states "*where an information society service is provided that consist of the transmission in a communication network of information provided by a recipient of the service, or the provision of access to a communication network, the service provider is not liable for the information transmitted, on the condition that the provider: a) does not initiate the transmission; b) does not select the receiver of the transmission; c) does not select or modify the information contained in the transmission. The acts of transmission and of provision of access include the automatic, intermediate and transient storage of the information transmitted in so far as this takes place for the sole purpose of carrying out the transmission in the communication network, and provided that the information is not stored for any period longer than is reasonably necessary for the transmission.*"

Cacheing activity is take in consideration by article 15 which affirms "*where an information society service is provided that consists of the transmission in a communication network of information provided by a recipient of the service, the service provider is not liable for the automatic, intermediate and temporary storage of that information, performed for the sole purpose of making more efficient the information's onward transmission to other recipients of the service upon their request, on condition that: (a) the provider does not modify the information; (b) the provider complies with conditions on access to the information; (c) the provider complies with rules regarding the updating of the information, specified in a manner widely recognised and used by industry; (d) the provider does not interfere, with the lawful use of technology, widely recognised and used by industry, to obtain data on the use of the information; (e) the provider acts expeditiously to remove or to disable access to the information it has stored upon obtaining actual knowledge of the fact that the information at the initial source of the transmission has been removed from the network, or access to it has been disabled, or that a court or an administrative authority has ordered such removal or disablement.*"

Finally, according to article 16, concerning host providers, an information society service provider is no liable for the information stored by a recipient of the service, on the condition that "*(a) the provider does not have actual knowledge of the fact that such activity or information is illicit and, as regards claims for damages, is not aware of facts or circumstances from which the illicit activity or information is apparent; (b) upon obtaining such knowledge or awareness through the communication of the competent judicial or administrative authority, acts expeditiously to remove or to disable access to the information.*"

Moreover, according to article 17 of the same decree, there is no general duty of surveillance on providers of services of transmission, cacheing or hosting, to monitor the information that they transmit or store, nor a general obligation actively to seek facts or circumstances indicating illegal activity. The Italian legislator issued this rules in order to implement article 15 of the EU Directive on Electronic Commerce, which forbids EU member states to "*impose a general obligation on providers, when providing the services covered by Articles 12, 13 and 14, to monitor the information which they transmit or store, nor a general obligation actively to seek facts or circumstances indicating illegal activity*" (Sartor and Cunha 2010, 360–61).

However, this model does not concern data protection, given that the EU Directive explicitly states that the providers' limitation do not affect data protection. This exception is contained in recital 14 of the EU Directive and specified in article 1(5)(b) of the Directive. Article 1(2)(b) Legislative Decree n. 70/2003 confirms that "*issues relating to privacy and processing of personal data in the telecommunications sector do not fall into the scope of this decree.*"

THE DECISION OF THE TRIBUNAL OF MILAN AND THE DECISION OF THE COURT OF APPEALS OF MILAN

Brief Description of the Facts

The facts of the case indicate that a group of teenagers, in Turin, Italy, audio and visually recorded themselves while they insulted, belittled, injured and harassed a young autistic boy named Francesco Giovanni De Leon. The 190 second film shows one of the perpetrators saying, "*Salve siamo, dell'associazione Vivi Down, un nostro mongolo si è cagato addosso e mo' non sappiamo che minchia fare perchè l'odore di merda c'è entrato nelle narici.*" The statement is translated as follows: "Hello, we are from the association Vivi Down, and one of our mongoloids[5] has defecated on himself, and we do not know what the fuck to do because the smell of shit has entered the nose." The video was created in late of 2006. It was uploaded in Google video in September 2006 in the city of Turin. Wrongly, the media reported that the child suffered from Down syndrome, however the disability was autism (Mendez 2010).

The video was viewed by a high number of people (more than 5,000 downloads). At a certain point it was the most popular category of "*video divertenti*" (funny videos). Users of Google video posted various messages commenting on the video (starting on October 4), apparently some flagged the video as being inappropriate and some e-mailed Google requesting for the video to be removed. However, evidence exists only for a flagging on 5 November 5, 2006 and an email request on the following day (Google stated that it was unable to provide documentation of all comments and flaggings). On November 7, the Italian Postal Police, after a communication from a citizen, requested Google to take down the video, which was removed on the same day. Thus, the video had remained available for about two months after it was initially posted.

The posting of the video gave rise to three distinct lawsuits. The first concerned the four students having an active role in the video (the three abusers and the movie maker). They were identified and were condemned by the Tribunal of Turin with a one-year sentence (work in social service), for assault and criminal defamation. The second lawsuit concerns the teacher for failing to prevent the offense: she was sentenced to a six-month suspended jail sentence for co-participation in assault and criminal defamation. The third lawsuit concerns Google, namely its Italian partner company (Google Italy and its executives.

The prosecutor of Milan started criminal proceedings against four leading Google executives (David Drummond—former president of Google Italy, George De Los Reyes—former member of the board of Google Italy, Peter

Fleisher—Google Privacy Counsel for Europe and Arvind Desikan—head of the Google Videos Project in Europe). The charges against them were criminal defamation and violation of data protection rules. With regard to defamation the indictment was of "concorso in diffamazione aggravata" (co-participation in aggravated defamation), that is of contributing to the defamation of the disabled teenager. With regard to data protection the indictment was that Google Italy was processing personal data, and in particular health data, illicitly, for the purpose of making a profit (Sartor and Cunha 2010, 357–58).

The Decision of the Tribunal of Milan

The case was decided on February 24, 2010 by the Italian Judge Oscar Magi: all four Google executives were acquitted with regard to the charge of defamation, and three of them were sentenced to a six-months suspended jail sentence for violation of data protection law. Written opinion of the court was released on April 12, more than one month before the date it was expected.

Since the accused were absolved from a charge of defamation, it is sufficient to give only brief attention.

The Decision of the Court of Appeal of Milan

The appeal was decided on December 21, 2012: all four Google executives were acquitted with regard to the charge of defamation, and for violation of data protection law. Written opinion of the court was released February 27, 2013.[6]

In relation to charge (a) the Court of Appeal confirmed the appealed judgment. Indeed, the court affirmed that a prior control on the content of video would not be effective, because of the technical limitations of filtering tools, in particular of those used at the time of the facts.

In addition, the appellate court adopted the same reconstruction of the court of first instance that qualifies Google Video as active hoster. Active hoster provides his input into the actual content or participates in the editing of the data: for all these reasons should be excluded that Google Video could continue to insist in its expected neutrality.

However, concerning charge (b), the Court affirmed that the criminal offence of unlawful data processing, stated by Article 167 of the Italian Personal Data Protection Code does not exist in the case under examination.

Indeed, according to Article 167, the crime exists only if three circumstances occur simultaneously: (a) deceit with the clear intention of either gaining profit or of causing damage; (b) data are processed violating one of the articles mentioned in article 167, and (c) data processing has in fact caused damage.

According to the court, the third circumstance has occurred, while the other two have not. Firstly, clear deceit has not occurred: in fact, it implies not only the purpose of gaining profit through ordinary commercial activity, but specific profit directly achieved by the accused. The court stated that the entrepreneurial characteristics of Google does not confirm evidence of specific intent of profit. Indeed there is not compatibility between willful misconduct (*dolo eventuale*)—which the Tribunal of Milan recognizes in the carelessness in company policies relating to privacy treatment—and the specific intent (*dolo specifico*) required by law.[7]

Secondly, data have been processed without giving data information to the data subject, but article 13 of the Italian Personal Data Protection Code, which affirms this duty, is not mentioned in article 167. Furthermore, no provision requires to ISPs to inform the user about existence and content of the code on personal data protection (Silvestre 2013, 1582).

Moreover, the duty to give data information to the data subject is the duty of the data processor, who in this case is the uploader of the video.[8]

Conclusion

The analyzed decisions state that it is impossible, because of the technical limitations of filtering tools, to control uploaded materials. However, liabilities in civil, criminal, and administrative fields could be identified both for YouTube and its executives.

On the one hand, the limitation on liability provided for by the European legislation is not applicable to: (a) the field of taxation; (b) questions relating to information society services with regard to the processing of personal data and the protection of privacy in the telecommunication sector; (c) questions relating to agreements or practices governed by cartel law; (d) specified activities of information society services (activities of notaries or equivalent professions to the extent involving a direct and specific connection with the exercise of public authority; representation of a client and defence of his interests before the courts; gambling activities which involve wagering a stake with monetary value in games of chance, including lotteries and betting transactions). In all these cases there is civil, criminal, and administrative liability.

On the other hand, as explained by the Court of Appeal of Milan, the exclusion of liability on the part of societies profiting from the offered service is not acceptable. Moreover, the analysis of Google's privacy policy underlines the existence and persistence of serious gaps. Although the Tribunal of Milan has decided that Google and its employees have to advise contents providers that they had to request permission from third parties to post material protected by the privacy code, Google's privacy policy has not changed.

Legal system aims to make responsible Internet providers, not by imposing on them an obligation to prior and general control of the content uploaded on the net, but by requiring them a duty of activation in presence of data collected unlawfully or illegally disseminated, in the presence of a specific request (Resta 2013, 1599).

In this same direction moved the Italian legislator: with the decree of 14/08/2013 n. 93 was expanded corporate criminal liability even in case of violations related to privacy.

NOTES

1. Directive 95/46/EC of the European Parliament and the Council of October 24, 1995 on the protection of individuals with regard to be processing of personal data and on the free movement of such data, http://eur-lex.europa.eu/LexUriServ/LexUriServ.do?uri=CELEX:31995L0046:en:HTML
2. Directive 2002/58/EC of the European Parliament and of the Council of July 12, 2002 concerning the processing of personal data and the protection of privacy in the electronic communications sector (Directive on privacy and electronic communications), http://eur-lex.europa.eu/LexUriServ/LexUriServ.do?uri=CELEX:32002L0058:en:HTML
3. Directive 2000/31/EC of the European Parliament and of the Council of June 8, 2000 on certain legal aspects of information society services, in particolar electronic commerce, in the Internal Market (Directive on electronic commerce), http://eur-lex.europa.eu/LexUriServ/LexUriServ.do?uri=OJ:L:2000:178:0001:0016:EN:PDF
4. Commission of the European Communities, Report from the Commission to the European Parialment, the Council and the European Economic and Social Commitee, First report on the application of Directive 2003/31/EC of the European Parliament and of the Council of June 8, 2000 on certain legal aspects of information society services, in particular electronic commerce, in the Internal Market (Directive on electronic commerce), http://eur-lex.europa.eu/LexUriServ/LexUriServ.do?uri=COM:2003:0702:FIN:EN:PDF
5. Derogatory term used for people affected by Down syndrome. The term is widely used in Romance language-speaking countries.
6. Corte Appello Milano n. 8611/2012 of December 21, 2012. http://www.oppic.it/index2.php?option=com_docman&task=doc_view&gid=487&Itemid=60
7. Alex Ingrassia, (2013, March 4). "La Corte d'Appello assolve i manager di Google anche dall'accusa di illecito trattamento dei dati personali." *Diritto Penale Contemporaneo* 2013, March 4), http://www.penalecontemporaneo.it/tipologia/0-/-/-/2115-la_corte_d___appello_assolve_i_manager_di_google_anche_dall___accusa_di_illecito_trattamento_dei_dati_personali/
8. Giusella Finocchiaro, March 5, 2013, commento on Law and the Internet, "Google vs. Vividown: the Milan Court of Appeal published the reasoning behing the decision," *The Finocchiaro Law Firm's blog*, accessed August 20, 2013, from: http://

www.blogstudiolegalefinocchiaro.com/wordpress/2013/03/google-vs-vividown-the-milan-court-of-appeal-published-the-reasoning-behind-the-decision/

REFERENCES

Alex Ingrassia. 2013. "La Corte d'Appello assolve i manager di Google anche dall'accusa di illecito trattamento dei dati personali," *Diritto Penale Contemporaneo*, 4 March, http://www.penalecontemporaneo.it/tipologia/0-/-/-/2115-la_corte_d___appello_assolve_i_manager_di_google_anche_dall___accusa_di_illecito_trattamento_dei_dati_personali/

Alexa Corporation. 2013. "The top 500 sites on the web." http://www.alexa.com/topsites

Boyd, Danah M. and Nicole B. Ellison. 2013. "Social network sites: Definition, history, and scholarship," *Journal of Computer-Mediated Communication* 13, doi: 10.1111/j.1083-6101.2007.00393

Cunha, Mario Viola De Azevedo and Lucia Marin and Giovanni Sartor. 2012. "Peer-to-peer privacy violations and ISP liability: data protection in the user-generated web," *International Journal of Law* 2 (2). doi: 10.1093/idpl/ips001

Di Ciommo, Francesco. 2010. "Programmi-filtro e criteri di imputazione/esonero della responsabilità on-line. A proposito della sentenza Google/Vivi Down," *Il Diritto dell'Informazione e dell'Informatica* 6, 829–57.

Finocchiaro, Giusella. 2013. "Google vs. Vividown: the Milan Court of Appeal published the reasoning behing the decision," *The Finocchiaro Law Firm's blog,* March 5. http://www.blogstudiolegalefinocchiaro.com/wordpress/2013/03/google-vs-vividown-the-milan-court-of-appeal-published-the-reasoning-behind-the-decision/

Hosch, William L. 2013. *Encyclopaedia Britannica.* 8th ed., s.v. "Youtube." Chicago: Encyclopaedia Britannica. http://www.britannica.com/EBchecked/topic/1262578/YouTube_

Kessler, Frank and Mirko T. Schäfer. 2009. "Navigating YouTube: Constitutina a hybrid information management system," in *The YouTube Reader*, ed. Pelle Snickars and Patrick Vonderau, 275–91. Stockholm: National Library of Sweden.

Mendez, Raul. 2010. "Responsibility for privacy violations in user generated content providers (Google case in Italy)." Master Thesis, University of Stockholm.

Mendez, Raul. 2011. "Google case in Italy," *International Data Privacy Law* 1 (2), 137–39.

Pablo Ameigeiras et al. 2013. "Analysis and Modelling of YouTube Traffic." *Transactions on Emerging Telecommunications Technologies* 23. doi: 10.1002/ett.2546

Pasqualino Silvestre. 2013. "La sempreverde tentazione di sostituirsi al legislatore," *Giurisprudenza di Merito*, 7/8, 1582.

Popoli, Anna Rita. 2012. "Cookies and Electronic commerce: a survey about actual knowledge of the issues concerning privasy" 10, 8–10. doi.org/10.2139/ssrn.2187496

Resta, Federica. 2013. "Diritti individuali e libertà nella rete nel caso Vivi Down," *Giurisprudenza di Merito* 7/8, 1599.

Rotman, Dana and Jennifer Preece. 2010. "The 'WeTube' in YouTube—creating an online community through video sharing," *Int. J. Web Based Communities* 6 (3), 318–20. doi: 10.1504/IJWBC.2010.033755

Sartor, Giovanni and Mario Viola de Azevedo Cunha. 2010. "The Italian google-case: privacy, freedom of speech and responsibility of providers for user-generated content," *International Journal of Law and Information Technology* 18 (4), 374. doi: 10.1093/ijlit/eaq010

Song, Seagull Haiyan. 2010. "A comparative copyright analysis of ISP liability in china versus the United States and Europe." *The Computer and Internet Lawyer* 27 (7), 1–18.

Suerz, Monica. 2013. "Internet tra diritto e giurisprudenza." *Tigor. Rivista di scienze della comunicazione e di argomentazione giuridica* 5 (1), 117–27.

Wasko, Janet and Mary Erickson. 2009. "The political economy of YouTube." In *The YouTube Reader*, ed. Pelle Snickars and Patrick Vonderau, 373–74. Stockholm: National Library of Sweden.

Wikipedia. 2013. "YouTube," https://en.wikipedia.org/wiki/YouTube

Chapter 13

Social Media in Education

Main Sources for Inclusion and Collaborative Learning

Magda Pischetola

The sociological discussion about the relationship between technology and society in the new millennium focuses more and more on the social and cultural impact of information and communication technologies (ICT). Internet represents a revolution not only from the point of view of the possibilities to communicate interactively and instantly but, more important, from the point of view of users' creativity in the production (and re-elaboration) of the contents. In this regard, the literature has investigated participatory culture (Jenkins 2006), network sharing (Rheingold 2008), and remixed productions (Lessig 2009) that generate new forms of cultural and social practices. After the popularization of the "network society" concept (Castells 2004), sociology has focused on the new science of networks (Benkler 2006; Franco 2011; Watts 2003). These studies point at changes in power distribution from a concentrated model toward a more decentralized one. They consider digital media as the *locus* where people meet, exchange ideas, share information, and, eventually, participate politically (Bennett and Segerberg 2013). In the online space, the social capital theorized by Bourdieu (1986) happens to have the same attributes—if not the same meaning—as social network. That is how, in the sociological discussion, digital inclusion overlaps the concept of social inclusion and impacts the field of education (Franco 2011; Helsper 2008; Neckerman 2004; Van Dijk 2005; Warschauer 2003).

The paper proposes an analysis of the relationship between technology and youth social practices through the lenses of education. It presents three comparative case studies, conducted in Italy, Ethiopia, and Brazil between 2009 and 2012, on digital inclusion programs which aim at fostering primary schools' innovation through one-to-one technologies. The results are used to discuss youngsters' learning practices and the ICT potential impact on digital and social inclusion.

The text is organized as follows: part one presents some issues regarding the impact of Web-based and mobile technologies on contemporary cultural and social practices; part two reflects on how social media changed the meaning of literacy and turned learning into an interactive process; part three questions the relationship between digital media and social inclusion; part four states the potential role of one-to-one technologies for education, presenting a comparative fieldwork research held in Italy, Ethiopia, and Brazil.

SOCIAL MEDIA AND CULTURAL CHANGE

At the end of the second millennium a number of social, economic, and cultural radical changes came together to give rise to a new form of social order: the network society (Castells 2004). The digital technologies are at the core of this transformation on a global scale, and force us to review some previous beliefs, in order to understand the new transitions.

First, the Internet has altered the idea that the information is difficult to find or takes long time to be transmitted. In fact, the global change has been characterized by the centrality of information as a primary good (Pischetola 2011; Van Dijk 2005). What we see now is a constant access to "the cloud": no longer do we carry information with ourself, on a CD or a pen drive, the only thing we need is some device to connect to the Internet. But to how many "clouds" do we access at the same time? And do we really separate work from entertainment? (DiMaggio et al. 2001). Not only we access online contents but we also read feedbacks, tags, comments, and recommendations that constitute an "ecosystem of reputation" (Lessig 2009, 61), based on tools that measure contents' and peoples' popularity and visibility, besides their relevance and importance. Thus, the Web space appears to be critical from the point of view of a clear understanding and elaboration of a message. That is, the abundance of information does not necessarily mean wealth of information (Kallinikos 2006). Instead, it calls for new skills and competences that a citizen of the twenty-first century should develop, including selection, retrieval, and processing information (European Commission 2008). Nowadays, it is not so important to access information, as much as it is to be able to create links among contents, and to overcome the fragmentation that the hypertext provides. If the ability to construct information from multimedia tools is lacking, then however good the operational skills are, information will not be capitalized on effectively (Bawden 2008).

Second, the advent of social media increased the possibility of communication and exchange of opinions, and it fostered users' participation to worldwide events—theoretically, whoever can publish their thoughts in a social network, share their notions on Wikipedia, and contribute to create what de

Kerckhove (1997) has named "connective intelligence." Following this idea, researchers have shown an increasing interest in how the use of social media modifies our relationship to knowledge. The past decade has seen the development of a "reticular" model of knowledge, where individuals belonging to different cultures meet and work together through social media (Brown and Duguid 2000).

In the late twentieth and early twenty-first century, sociologists like Castells (1996) and Watts (2003) popularized the view of networks as spaces of interaction, communication, and social organization. "When solving complex problems in ambiguous environments, individuals compensate for their limited knowledge of the interdependencies between their various tasks and for their uncertainty about the future by exchanging information—knowledge, advice, expertise, and resources—with other problem-solvers within the same organization" (Watts 2003, 273). The network is a characteristic treat of human social interaction. The social networking represents a key strategy for information sharing, the formation of a political and cultural identity, the search for mutual support, and the sense of belonging to a social group. Through the virtual social networks, space and time get tremendously reduced, while platforms for social interaction are created for varied purposes.

The networks begin to form lines of communication that are decentralized rather than centralized, and generate unpredictable flows of information and collective self-organization. "At the other end of the spectrum of social ties, we see new platforms emerging to generate the kinds of bridging relations that were so central to the identification of 'weak ties' in social capital literature. Weak ties are described in the social capital literature as allowing people to transmit information across social networks about available opportunities and resources, as well as provide at least a limited form of vouching for others—as one introduces a friend to a friend of a friend. What we are seeing on the Net is an increase in the platforms developed to allow people to create these kinds of weak ties based on an interest or practice" (Benkler 2006).

Why is this relevant to understand the influence of social media in education? Because innovation depends more and more on groups of people working together, enhancing their diverse talents and specialties, and solving multifaceted interdisciplinary challenges (Lindegaard 2010). Nevertheless, while practices of sharing increase the potential of dialogue and the exchange of ideas, at the same time they represent a major challenge to the transformation of information into knowledge. In fact, as we will examine throughout this paper, the peculiarity of the contemporary society is not the importance of information itself, but the creativity and inspiration that the information receivers exercise when using it.

Third, the virtual sharing practices have led us to reconsider the meaning of authorship, and see it as a collective process of "remix" (Lessig 2009), through which contents—texts, images, videos, music—are constantly reinterpreted and re-produced. The social platforms of Web 2.0 offer users the possibility to create new media, which represents the start of a potential reshaping of the cultural production enterprise. In fact, the traditional gatekeepers of culture—journalists, editors, producers, media distributors—have all been forced to adjust to the realities of participatory cultural production. "The newspaper editor, the news station, the radio host, who is able to promote and filter based on certain viewpoints (political, philosophical), is losing ground to the individual. Individuals can form and create their own spaces and networks of news, information, and knowledge filtering and dissemination" (Siemens 2006, 23). As creative productions on digital platforms grow, not only they change business, they also change how we think about culture as a business. On one hand, we are free to choose when to watch a TV program or a film, as we would choose when and where to read a book, according to our schedule and our wishes. On the other hand, we have to be aware that "to write in this medium is to know that anything one writes is open to debate" (Lessig 2009, 67). That is, using the facilities of a free digital network, anybody can share the writing process, mixing all kinds of contents, and creating what can be described as a large-scale "collage." These are often peer-to-peer production practices, a term that refers to "production systems that depend on individual action that is self-selected and decentralized, rather than hierarchically assigned" (Benkler 2006, 28). As Benkler points out, a very interesting aspect of these kinds of practices are the actions based on social norms performed by millions of users, who work together but at the same time write, play, and create entirely independently from each other. In 2000, Rifkin had already foreseen how these social changes would have an impact on the market system. Commenting on the first peer-to-peer online service of music exchange, named Napster, the author wrote:

> while there has been a great deal of angst over how to protect copyright and ensure that musicians and music companies are rewarded for their contribution, little attention has been paid to where the whole network process is leading society. Napster speaks to a much deeper change in the economy. In the coming century, short-term access between servers and clients operating in networks is going to redefine our social dynamics as powerfully as did exchange of property in markets in the past century. Today, Napster is an oddity. In the future, it will be the norm.

Now, remixing and peer-production practices are important for education because they foster an interest-based learning and become an organic part of the process by which children develop cultural literacy (Jenkins 2006). In conclusion, a central aspect of digital revolution is the possibility for the users to become active participants in cultural production. The next chapter will discuss the implications this trend has for education.

IMPLICATIONS FOR EDUCATION: TOWARDS NEW CONCEPTS OF LITERACY AND LEARNING

These assumptions lead us to recognize the evolution of the ways people access information and transform it into knowledge in a sharing mode. This has two kinds of implications for schooling. On one hand, it implies considering a broader definition of literacy, which "is plural, dynamic and situational, relating not only to basic writing and numeracy skills in one language, but also the ability to identify, understand, create, communicate and compute information in various languages" (UNESCO 2010). It also means that literacy involves posing problems that will develop cognitive abilities—learning how to identify key concepts, make connections between multiple ideas, ask pertinent questions, identify fallacies, and formulate a response (Thoman and Jolls 2005). These skills are at the foundation of intellectual capital, and prove to be necessary for exercising full citizenship in the network society. "To acquire competences, learners need, from an early age, to 'learn to learn' by reflecting critically on their learning aims, managing their learning with self-discipline, working autonomously and collaboratively, seeking information and support when necessary, and using all the opportunities of new technologies" (European Commission 2008, 6).

On the other hand, participating in social construction of knowledge through the networks and being able to produce contents creatively are part of what are now considered competences for the contemporary world (Bawden 2008). Therefore, formal education must consider the way that pupils learn outside of the school, and try to integrate the sharing practices that youth perform while using the media into daily school activities. This means reconsider the learning paradigm, towards a child-centered and collaborative learning model.

The notion of collaborative learning is not new: it was originally described by the socio-constructivist perspectives of the twentieth century. Vygotsky (1978) theorizes collaborative learning and points out the need to provide a space for interaction between student and teacher, where learning occurs. He calls this the "Zone of Proximal Development." By introducing this concept, the author analyzes the development of skills from the interaction with others and concludes that the construction of knowledge is an interactive process. In this perspective, learning should be considered a social activity. This requires not only the development of the individual, but also the presence of the community to which he belongs. Stating that all knowledge is the result of an active construction endorsed by the subject, the socioconstructivists claim that "learning is not the result of development, learning *is* development" (Fosnot 2005).

With the increasing presence of networks in all spheres of society, new learning theories arise and new interpretations regarding the social construction of knowledge. Most of them reflect, implicitly or explicitly, the influence of Vygotsky's theory about cognition as a product of a particular social environment. Henderson (1996), for example, points out that social learning affects three areas of knowledge: technical skills, academic skills, and cultural background of each individual. These three areas overlap in "Zones of Proximal Development." Salomon (1993) also states the existence of social construction of knowledge. For the author the cognitive elaboration of information occurs in conjunction with the other individuals, because of the possibilities that constant interactions between people open up.

These theoretical approaches highlight not only that networks established through relationships improve the exchange of information, but also that they allow one to share knowledge and develop creativity. Following this idea, one can affirm that access to the information provided by digital technologies is the first step in the construction of knowledge. However, research indicates that virtual networks can also be inconsistent from the point of view of learning, when they are used in a traditional way (Franco 2008; Pischetola 2011; Warschauer 2003). Therefore, to avoid falling into technological determinism, it is crucial to recognize that it is people's relationships which determine behavior, not the technology itself (Smith and Marx 1994).

In his famous work, "Society without Schools" (1970), the Austrian educator Ivan Illich has imagined the possibility of a worldwide network of computers, and that students become independent of the teacher in access to information. In the last decades, an increasing number of projects have addressed school innovation through technology as a driver for change. Many of them have already failed, indicating that a deterministic approach to innovation will not change the school (Pischetola 2014; Warschauer 2003). The aim of this paper is to go one step further by relating educational change to investments in human resources and capacity building, rather than technology.

INCLUSION THROUGH THE SOCIAL MEDIA

Accepting the premise that ICT, and especially the Internet, has changed the landscape of information access, we recognize that in contemporary societies it is fundamental to have access to digital resources, in order to be "socially included" (Castells 2004). The availability of a huge amount of data provided by social media and the dialogue with the other peoples contribute not only to the change of the individual way of thinking, but also in the participation to the "public sphere" (Habermas 1989). Based on these assumptions, many authors argue that lack of access to information and incompetence in its use limit citizens' democratic participation (Giddens 2000; Norris 2001; Nussbaum 1998; Splichal and Wasko 1993). Thus conceived, information could be among what Rawls (1976) identified as primary goods, which are basically things that every rational man is presumed to want, because they have a use whatever a person's rational plan of life. In this perspective, access to information becomes the first step to knowledge building, and unequal access becomes a core reason for social inequality.

But we must also recognize the existence of a second level of inequality, which mainly depends on what people *do* with the information they have access to (DiMaggio et al. 2004). Bucy and Newhagen (2004) contend that the usual identified characteristics of a population who has or has not access to ICT (e.g., race, income, gender, social status, etc.) only explain some aspects of the matter. To picture a more complete landscape, these characteristics should be expanded to include psychological, cultural, and sociological differences, which have an influence on the way people meet technology and use it. These reflections have two main implications for the definition of digital inclusion.

Digital Inclusion is about Human Capital

In terms of literacy and learning, not only ICT access is required, but *meaningful* access. This implies considering whether an individual has the necessary means to access the technology not only in terms of affordability or infrastructure, but also with abilities and cognitive competences. Education should encourage the ability to become interpreters, together with the awareness of cultural difference and the pursuit of mutual respect (Nussbaum 1998). In this perspective, social media can be used to better respond to the individual needs of students. They can provide, through teachers' mediation, a meaningful access to information, as well as the development of pupils' critical potential, procedure autonomy, and creativity (Pischetola 2011). The familiarity with the technology and confidence with the Web promote an easier access to contents. Those who have higher levels of education adapt more easily to technological innovations, in a virtuous circle which is the base of economic and human development. The level of education provides, in fact, that initial cognitive advantage that leads users to process the new information more effectively, so the return on investment in knowledge will be greater (Norris 2001). By consequence, if we are to promote social development through technology, we must reconsider the meaning of access to ICT from the point of view of the capacity to use digital technologies. Stressing the importance of human capital in relation to digital technologies means to consider, firstly, the access to cognitive skills. These are, in fact, the skills necessary to learn and succeed in a complex learning environment, as well as the base of equal educational opportunities.

Digital Inclusion is about Social Capital

A very important component of digital inclusion is social support, which makes the individual feel confident to make effective use of technologies and to become an active participant in the Web 2.0 platforms. The literature on the diffusion of innovations has always emphasized the importance of social support networks in the spread of new technologies (Rogers 2003). Those who are more exposed to innovations in their surroundings are more likely to adopt new technologies. This is due to several factors. First, the exchange of information existing in a social context provokes one's initial involvement, through personal motivation (Van Dijk 2005). Second, having friends and family who also use technology may provide technical assistance and psychological encouragement (DiMaggio et al. 2004). Third, in a macro perspective, the capital of a community can optimize development programs and human resources through widespread partnerships and synergies. Beyond the strategic alliances between institutions, whose political role is certainly essential, it is necessary to identify initiatives and mechanisms which improve local knowledge and answers to the needs, for example with the development of ad hoc software or content translation in the local language (Pischetola 2011). In all these situations, the understanding of the social constellation arises when we become aware of the fact that the social potential belongs not to the people, but to the network itself (Franco 2011). This is to say that if a network is composed by a set of spots and lines, where the first ones are the people and the second ones the relations, there is no doubt the lines represent the most important element of the picture (Martinho 2003). In this sense, the theorization of Bourdieu (1986) of social capital proves to be very actual, when we see that being linked to a network of relationships or being members of a group represent useful resources for mutual help and, therefore, an important result of energy investment.

These considerations lead us to revise the meaning of social exclusion in a broader sense, considering not only the traditional factors of poverty, related to income, age, social position, but also aspects partially related to access to social media and the Internet, such as the chances for employment and education, the opportunities for political and civic participation, as well as the access to information (Helsper 2008; Van Dijk 2005). Integrating a broader concept of literacy and conceiving new forms of learning are the key factors for changing education, toward a paradigm that contemplates digital inclusion as a key for social inclusion (Pischetola 2011).

ONE-TO-ONE PROGRAMS AND SCHOOL INNOVATION

One-to-one technologies are low-cost portable laptops, notebooks, or tablet PCs for teachers and school-aged children, to be used both in the classroom and at home. Launched in 2005 by the Massachusetts Institute of Technology, the *One Laptop Per Child* (OLPC) (http://one.laptop.org) program was the first and largest attempt in the world to promote school innovation through such devices. Using different technologies, many other projects reproduced the OLPC pedagogical formula around the world. One of them is the Brazilian *Um Computador por Aluno* (UCA), which besides the use of a different laptop and operational system, is the OLPC exact reproduction.

A key pedagogical thought behind the design of the OLPC and the UCA projects is the activist educational thinking, which argues that the most effective way of learning is personal discovery (Papert 1993). Students can learn by doing, while the teacher's role is to mentor the learner, as a facilitator of the whole process. The idea is inspired by Dewey's approach (1944), known as "learning by discovery," and based on the concepts of education as reconstruction of experience and motivation as the driving force of learning. This way of learning involves reasoning procedures which are very different from passive reception of contents, as the student needs to reorganize the new information in order to integrate it into the cognitive baggage which already belongs to him. Another fundamental source for the one-to-one approach is constructivism (Vygotsky 1978): social support allows children to accomplish with assistance what they could not do on their own, and, this way, learn how to accomplish the task independently. On these grounds, every child who has access to a mobile digital device is expected to become an active participant in a learning community, and contribute to a substantial change of the traditional school environment.

Summarizing, the basic concept of the OLPC and UCA initiatives implies that

1. Every child owns a laptop, to be used as school and at home. The laptop is a private and personalized property, as opposed to the usual models where ICT is shared among students;
2. The laptop is used in the classroom transversal to all disciplines, as opposed to the concept of the class of Informatics;
3. The child is not a passive consumer of knowledge, but an active participant in a learning community.

According to constructivists, the process of sharing other persons' points of view results in learners building understanding together that would not be achievable alone (Brown and Campione 1994). Transferred to the one-to-one programs, the model is that of the community of learners in which each member is at the same time apprentice and teacher.

The following paragraph introduces a comparative research on three case studies. The first two were conducted in Italian and Ethiopian primary schools, where OLPC was implemented, during the school year 2008/9. The third one was carried out in Brazil with the same methodology, on the parallel UCA program, in 2012. The research question that drives them is whether a one-to-one technology approach can bring about an educational change in terms of digital literacy and collaborative learning.

RESEARCH METHODOLOGY

The comparative research followed a qualitative methodology, based on three different tools: participant observation in the classrooms throughout the school year (ten/country); focus groups with children at the end of the school year (ten/country); intensive interviews with teachers and coordinators throughout the school year (twenty/country).

The overall data collection involved thirty classrooms, and led to a total of thirty focus groups with the pupils and sixty interviews with the teachers/coordinators. The initial hypothesis is that the development of children's skills related to digital literacy could result in a change in the learning environment towards a more collaborative model.

The research started from Van Dijk's concept (2005) of *digital skills*, which comprise complex factors required to access and work with digital technologies. The author identifies three kinds of digital skills:

- Operational skills which correspond to the general technical ability to work with hardware and software;
- Informational skills which are needed in order to make a start on the real handling of information and knowledge. Among these skills there is the ability to work with hyperlink structures and, more substantial, the ability to find, select, process, and evaluate information in specific contexts;
- Strategic skills, which can be defined as "the capacities to use computer and network sources as the means for particular goals and for the general goal of improving one's position in society" (Van Dijk 2005).

The most popular view is that access to technology is certain when the operational skills are mastered. However, research of the last decade has called attention to other forms of competence, such as the effective understanding of the retrieved information, and the strategic use of it for individual purposes (DiMaggio et al. 2004; EU 2008; Pischetola 2011).

Given these premises, the study takes into account the possibility that children engaged in laptop immersion programs are more likely to learn how to learn, in a problem-solving perspective. It considers creative production with a broader meaning, which includes the *use* of the devices made available with OLPC and UCA projects (Alvermann 2002). Finally, it analyzes what are the skills that improved significantly among students and teachers, and tries to find meaningful correlations with the setting changes that were detected by teachers during the school year.

MAIN FINDINGS

The method used for data analysis is the one suggested by the authors of Grounded Theory (Glaser and Strauss 1967). The transcribed texts are read with an inductive approach, in which individual events, situations, and personal opinions are grouped to form general categories of interpretation. In all three case studies, the findings suggest that technology certainly does have a role in encouraging discovery learning and sharing of knowledge. Nevertheless, the contexts present interesting differences in the results.

ITALY: CHILDREN SKILLED IN INFORMATIONAL RETRIEVAL AND PROBLEM SOLVING

In Italy, the study shows the emergence of a problem-solving culture. According to the teachers' opinions, the laptop especially helps improve information retrieval and elaboration, as well as motivation to discover new features and solutions. Most of the interviewed teachers recognize that working through a discovery process allows students to develop their problem-solving skills, and being more autonomous in the completion of a task (Mangiatordi and Pischetola 2010).

The ability to manage a new tool and the greater self-esteem children gained seem to have an effect on their general behavior towards class mates, especially in promoting peer-to-peer support as a common learning practice. As one teacher says about her students,

> From being students they became teachers. They had to give the tool instructions to make it work the way they wanted and, by doing this, they acquired security and self-esteem. They also started to collaborate . . . they didn't so much before, because the exercise book was very personal. They were used to protect it from the mates' look, to avoid cheating. Now there's a great collaboration. They enter in the mesh network (*intranet*), chat, and create groups of friends. The concept of "cheating," even the word itself, does not exist anymore. Cause they just help each other. I also ask them to consult each other with some

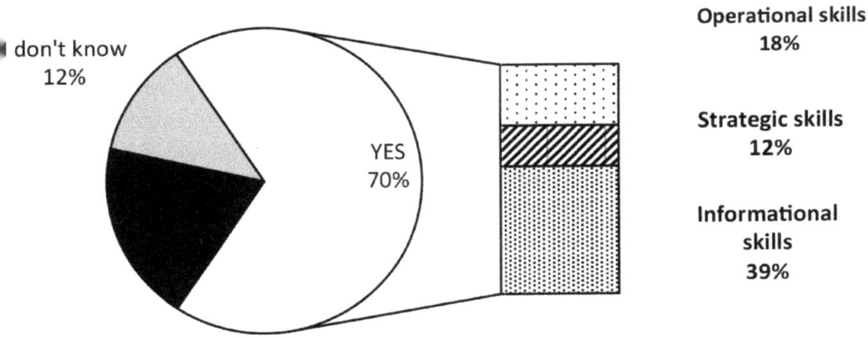

Figure 13.1

> of the uses. There's a child, for example, who's got better competence and his mates ask him, but I also do sometimes! (Teacher, third grade).

Many teachers notice how this change in the collaboration dynamics inside the classroom has a positive impact on social inclusion, both concerning students with cognitive problems and the children of immigrants (who represent 17% of the sample).

> I would never have expected two children of my class to do so well in this project. I mean, it's children who've always had the role of "followers." 'Cause it was always someone else who was the leader, who knew better how to communicate, etcetera. But now for example, one of these children is a lightning: he goes everywhere, he knows a lot, and he manages the group! (Teacher, fourth grade).

> Since the project started I have seen how this machine has been positive for children with problems. If you're able to motivate them . . . well, they become teachers of the other mates! Their self esteem increases, because if they can succeed in this case, they learn . . . it means they can in general (Teacher, fourth grade).

Teachers underline how the laptop helped socially excluded children—especially second generations of immigrants—to become the most respected "experts" in technology (Pischetola 2011).

ETHIOPIA: STUDENTS' PRACTICES OF A "REMIXED AUTHORSHIP"

In Ethiopia, the outcomes suggest that social support acts as a driver for change in learning practices. Ethiopian children employed effective search strategies to face the absence of the Internet connection, through the construction of an impressive social network among them. Compared with the other fieldwork studies, Ethiopian children proved themselves the most creative users of the laptop. The study outlined the great awareness pupils have about the potential uses of the laptop, much aside from the traditional use of it as a normal book. They became experts of the most difficult activities uploaded on the laptop, thanks to self-initiative, autonomous discovery, and social support among peers. Teachers recognize children's expertise, and define it through four different kinds of skills observed.

It must be noted that the OLPC laptop implements a proposed wireless standard that allows the computers of the same local area to communicate with each other without the Internet. This feature represents a very innovative feature for children, as it allows communication by chat and file-sharing. This characteristic partially replaces the Internet connection, which is not available and responds with many resources to the youngsters' thirst for knowledge. The researched schools turned into a laboratory of "remix," with the existing materials being constantly combined or edited to produce new product contents.

Some influences of technology on collaboration dynamics are also evident. As one child explains,

> *Child:* I like the chat very much. I do it in the classroom. My friend would be in a corner and I'd be in the opposite corner, so we can chat.

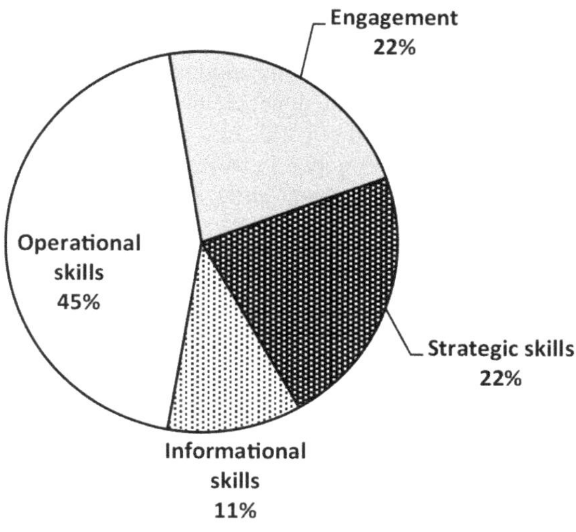

Figure 13.2

Researcher: Do you do it during lesson?

Child: No, during break time. Then, there is a game called Pacman. I play it with other friends, usually three of us. And then there is someone who wins and will be rewarded.

Researcher: So you have a sort of competition?

Child: Yes, and I hope they're losing [*laughs*]. This is another game, labyrinth. We also have a competition: you have to find out how to get to the end. The other activity is record activity: we record pictures and videos that we exchange between us.

These collaborative uses of the laptop—repeatedly mentioned in the focus groups—match the purpose of creating a community around the tool, based on sharing procedures, contents and products. The consequence of these new practices is a great motivation to discover new features of the laptop, and to exchange more information in the peer community. The enthusiasm for the features of the laptop is evident in the following answers given by children during the focus groups:

I saw this one in TV and I recorded it. I also recorded a video of this person, he's an actor, and the coach of the Arsenal team. Then . . . I was watching a movie, I saw a beautiful car and I recorded it . . . This is my area and I also recorded it from TV. When I'm bored I like to watch things again . . . that's why I record (Boy, third grade).

I've got two pictures, one is a greeting card because I took a picture of my brother's birthday. I've got another camera, but I prefer taking pictures with the laptop, because then I can edit them (Girl, fifth grade).

I will write something in the writing activity and I will combine it with a picture I took. I use Google to know more about people like Mandela and also to know more about languages spoken in the world. And also to know about different historical places that are out of Ethiopia. I can read the history. I use the write activity to write my biography, different cities and their name, these kind of things (Girl, sixth grade).

On the other hand, the assumptions concerning the nature of discovery learning intrinsic to the OLPC design generated some difficulties in the Ethiopian teachers' methodology, as the dominant mode of education places high importance on hierarchy and authority (Everts et al. 2007). The introduction of the laptop in itself did not change the traditional way of teaching, even though some of the teachers observed (23%) demonstrated more openness towards the new technology and tried a more interactive approach with students. It was clear from the interviews that integrating ICT in daily lessons means for teachers extra work, extra energy, and motivation. As teachers stick to the traditional way of teaching, children do not have other support than that provided by classmates. And they confirm through their remix practices the hypothesis that social support is the starting point of learning conceived as an interactive process.

BRAZIL: STUDENTS' COMMUNICATION ON SOCIAL PLATFORMS

In Brazil, the results indicate that students' performance did not improve significantly in terms of strategic use of the laptop. Many teachers underlined during the interview that the new technology has brought lack of attention and distraction in their classrooms, and that their students mainly improved in technical skills.

Nevertheless, the most striking result to emerge from the data is the very high level of engagement that teachers have detected in the researched classrooms, especially from the point of view of the social exchange and communication. Fieldwork observations have confirmed that children were engaged often in entertainment activities, such as online games and chats on social networks during the class time, and mostly without teachers' permission. Figure 13.3 shows that gaming and chatting are among the activities that the students say to prefer.

Teachers who complained about students' distraction and blamed the Web connection were also those who received the laptop with greater caution. Their manner of using computers in teaching has been the traditional way,

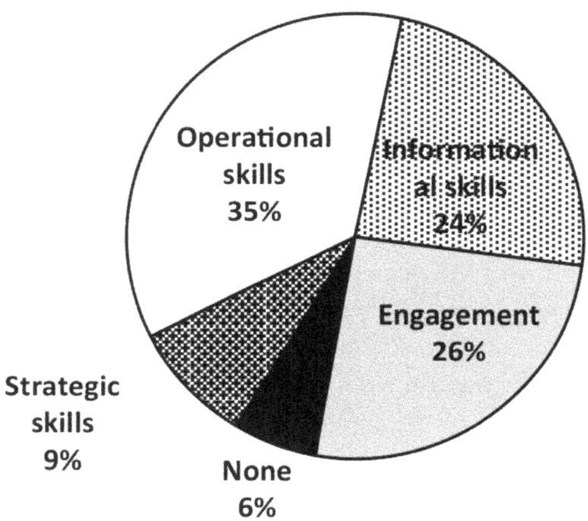

Figure 13.3

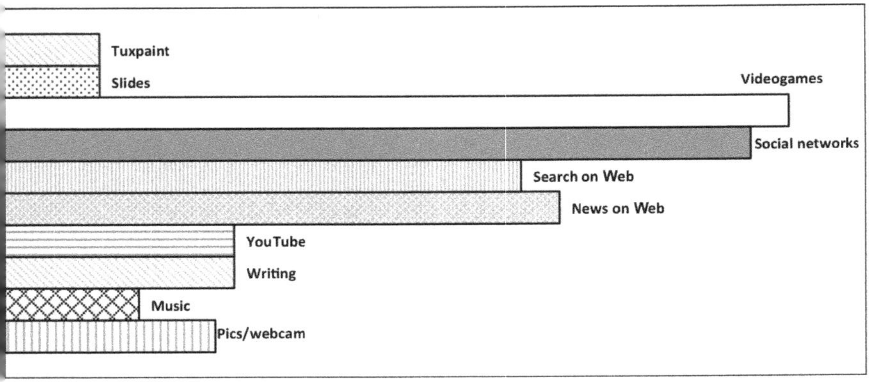

Figure 13.4

where the children were asked to follow the steps outlined by the teacher. This has prevented the children from learning by discovery and, in general, increased dispersion in the classroom. Whenever this approach did not change during the school year, a gradual loss of control and authority was observed among teachers, with a feeling of frustration and disappointment that turned into a negative perception of the project (Pischetola 2011).

One unanticipated and very interesting finding is that older students, who also received the laptop, imposed a self-regulating frame on themselves for the use of the Internet at school. Whenever the rules for the use of the laptop were not clear and led to distraction, teenagers reacted by building their own set of rules, basically looking for self-discipline.

> Here in my collage I put on the negative side all the icons of social networks, because this thing is destroying the studies. Many people don't listen during lessons, they're all addicted to games. I don't even bring the laptop at school, only to do some works on it. I had to repeat the year, this is my second time in 8th grade, but now I got it. I don't even access facebook anymore. It harmed me! (Boy, 8th grade).

> Everybody here spends their time on social networks. It seems sometimes that they understand the Internet as if it was only chatting on social networks! Nobody search for information that could increase their culture and intelligence (Girl, 13 years old).

On the other hand, distraction was not detected in school contexts where teachers approached the use of the laptop with a more task-oriented methodology (Pischetola, 2011). In these classrooms, pupils were encouraged not only to search for new information, but also to use their existing knowledge to build new skills. It goes without saying that a methodology of this kind fosters collaboration between students, joined in the search for a solution or in the advancement of a task. It is interesting to notice that in these classrooms students privileged virtual contact as a more interesting space to play and talk, almost as in a sort of parallel reality. As reported by one of the interviewed teachers,

> They interact with each other, especially through the games online. They're playing for example with the game named "Transformice," which they like so much, in the same room, the same physical space. But at the same time, they're in the same virtual space. I found it very interesting, the way they talk at each other. The mate is there, close by, and every now and then someone says loud: "don't take me! I won't let you!." They get completely involved. [...] This made them turn accomplices, you know? Some of them made friends with mates of a different class, thanks to this game. They create nicknames, and later they discover who is who in the school, and they become friends in name of a space they've shared elsewhere, not in the physical space. They would have never talked to each other if it wasn't for the experience they've shared online.

It is clear from these findings that operational and informational skills were comprehensively detected, in different proportions, in all the examined contexts. But it must also be noted that strategic skills, which include the ability to use information in a more complex way, as it can be for example remixing and producing contents, did not appear as a striking finding, with the exception of the Ethiopian context. Even there, though, digital literacy, as we have previously intended it, was not attained by many.

Another crucial point raised by the research is that the integration of a new technology represents an enormous challenge to teachers' established methodology. The one-to-one formula brings calls for a new learning paradigm, but it does not consider how teachers will coordinate the use of a new device in their often overcrowded classrooms without receiving any methodological assistance. That is to say that rather than being an innovative solution, one-to-one laptop programs may simply amplify what is already occurring, for better or worse, in schools.

CONCLUSIONS

The outcomes designed from the presented studies suggest that the one-to-one technology embodies an ethos, in which information and media literacy skills act as a main driver for change. In all the examined realities children were active users of the laptop. In the Italian context, they mainly developed a problem-solving ability, which is related to information retrieval and autonomous elaboration of the accessed contents. In Brazil, social networks and games have been considered by children useful platforms to extend their network of friends. Ethiopian classrooms have turned into workshops of authorship, where children have learned how to connect with each other even without the Internet, and have developed strategic skills to generate new contents with creativity.

As we discussed in this chapter, a strong relationship between digital literacy and collaboration has been reported in the literature. The findings corroborates this idea, indicating how collaborative dynamics among children occurred especially in the classrooms where pupils have developed a wider range of skills. In this sense, the Ethiopian context stands out for the significant collaborative practices that children were able to generate in their work of "remix" and original production. Engagement has had a very important role in these dynamics. Hence, it could conceivably be hypothesized that personal curiosity and motivation toward the use of the laptop is the element that makes the difference. Nevertheless, the findings suggest also that to accomplish the mission of innovating school practices, it is necessary to invest in methodological training for teachers. In fact, the evidence is that students are more than willing to comply with prescriptions given by teachers when the rules for the use of the laptop are clear and the work is task-oriented. Many are, consciously or unconsciously, working to bring about change in their learning results. They end up also changing their relationship with others, and thus experiencing social inclusion. In conclusion, the essential elements for a change of learning through social media seem to be the development of critical skills for information retrieval, and the ability to use real and virtual social networks as a space for exchanging opinions and peer-production. Technological skills are not enough, especially when the envisaged outcome is a cultural change of learning practices.

REFERENCES

Alvermann, D. E. 2002. *Adolescents and Literacies in a Digital World.* New York: Peter Lang.

Bawden, D. 2008. "Origins and Concepts of Digital Literacy." In *Digital Literacies: Concepts, Policies and Paradoxes,* ed. C. Lankshear and M. Knobel. New York: Peter Lang.

Benkler, Y. 2006. *The Wealth of Networks.* New Haven: Yale University Press.

Bennett, L. and A. Segerberg. 2013. *The Logic of Connective Action: Digital Media and the Personalization of Contentious Politics.* New York: Cambridge University Press.

Bourdieu, P. 1986. "The Forms of Capital." In *Handbook of Theory and Research for the Sociology of Education,* ed. J. Richardson. New York: Greenwood Press.

Brown, A. L. and J. Campione. 1994. "Guided Discovery in a Community of Learners." In *Classroom Lessons: Integrating Cognitive Theory and Classroom Practice,* ed. K. McGilly. Cambridge: MIT Press.

Brown, J. S. and P. Duguid. 2000. *The Social Life of Information*. Boston: Harvard Business School Press.
Bucy, E. P. and J. E Newhagen. 2004. *Media Access: Social and Psychological Dimensions of New Technology Use*. Mahwah: LEA.
Castells, M. 1996. T*he Rise of the Network Society, The Information Age: Economy, Society and Cultur*e. Cambridge: Blackwell.
Castells, M. 2004. *The Network Society: A Cross-Cultural Perspective*. Cheltenham, UK; Northampton: Edward Elgar.
Coleman, J. S. 1990. *Foundations of Social Theory*. Cambridge: Belknap Press.
De Kerckhove, D. 1997. *Connected Intelligence: The Arrival of the Web Society*. Toronto: Sommerville House.
DiMaggio P., E. Hargittai, W. Russel Neuman, and J. Robinson. 2001. "Social Implications of the Internet." *Annual Review of Sociology* 27.
DiMaggio, P., E. Hargittai, C. Celeste, and S. Shafer. 2004. "From Unequal Access to Differentiated Use: A literature Review and Agenda for Research on Digital Inequality." In *Social Inequality* ed. K. Neckerman. New York: Russell Sage Foundation.
European Commission. 2008. *Improving Competences for the 21st Century: An Agenda for European Cooperation on Schools' Communication*. Brussels: Commission of the European Communities.
Everts, B., M. Herren, and D. Hollow. 2007. *Ethiopia Implementation Report*. Addis Ababa: Eduvision.
Fosnot, C. T. 2005. *Constructivism: Theory, Perspectives and Practices*. New York and London: Columbia University.
Franco, A. 2008. "Escola de Redes: Tudo que é sustentável tem padrão de rede." *Sociedade do Conhecimento*, V. 2. Curitiba, PR: ARCA.
Franco, A. 2011. *A Rede. Escola de redes*. Série Fluzz, vol. 1.
Giddens, A. 2000. *Runaway World: How Globalization is Reshaping Our Lives*. New York: Routledge.
Glaser, B., and A. Strauss. 1967. *The Discovery of Grounded Theory*. New York: Aldine.
Habermas, J. 1989. *The Structural Transformation of the Public Sphere: An Inquiry into a Category of Bourgeois Society*. Cambridge: The MIT Press.
Helsper, E. J. 2008. *Digital Inclusion: An Analysis of Social Disadvantage and the Information Society*. London: OII.
Illich, I. 1970. *Deschooling Society*. Cuernavaca, Mexico: CIDOC.
Kallinikos, J. 2006. *The Consequences of Information*. Cheltenham-Northampton: Elgar.
Jenkins, H. 2006. *Convergence Culture: Where Old and New Media Collide*. New York: New York University Press.
Lessig, L. 2009. *Remix*. London: Penguin.
Lindegaard, S. 2010. *A Revolução da Inovação Aberta: A chave para a nova competitividade nos negócios*. São Paulo: Évora.
Mangiatordi, A. and M. Pischetola. 2010. "Sustainable Innovation Strategies in Education: OLPC case studies in Ethiopia and Uruguay." In *Organizational,*

Business, and Technological Aspects of the Knowledge Society, Communications in Computer and Information Science, ed. M. D. Lytras. Berlin: Springer.

Martinho, C. 2003. *Redes. Uma introdução às dinâmicas da conectividade e da auto-organização*. Brasília: WWF.

Norris, P. 2001. *Digital Divide? Civic Engagement, Information Poverty and the Internet Worldwide*. New York: Cambridge University Press.

Nussbaum, M. C. 1998. *Cultivating Humanity: A Classical Defense of Reform in Liberal Education*. Cambridge: Harvard University Press.

"One Laptop Per Child," http://one.laptop.org

Papert, S. 1993. *The Children's Machine, Rethinking School in the Age of the Computer*. New York: Basic Books.

Pischetola, M. 2011. *Educazione e divario digitale. Idee per il capacity building*. Milano: Unicopli.

Pischetola, M. 2014. "Teaching with Laptops: A Critical Assessment of One-to-one Technologies." In *Media and Education in the Digital Age. Concepts, Assessments, Subversions*, ed. M. Stocchetti. Frankfurt: Peter Lang.

Rawls, J. 1976. *A Theory of Justice*. Cambridge: Belknap Press.

Rheingold, H. 2008. "Using Participatory Media and Public Voice to Encourage Civic Engagement." In *Civic Life Online: Learning How Digital Media Can Engage Youth* edited by Bennett, L. Cambridge: The MIT Press.

Rifkin, J. 2000. "Where Napster Has Gone, Others Will Follow." *Los Angeles Times*, August 21.

Rogers, E. 2003. *Diffusion of Innovations*. New York: The Free Press.

Salomon, G. 1993. *Distributed Cognitions: Psychological and Educational Considerations*. Cambridge: Cambridge University Press.

Smith, M. R. and L. Marx. 1994. *Does Technology Drive History? The Dilemma of Technological Determinism*. Cambridge: MIT Press.

Splichal, S. and J. Wasko. 1993. *Communication and Democracy*. Norwood: Ablex publishing Corporation.

Thoman, E. and T. Jolls. 2005. *Literacy for the 21st Century: An Overview and Orientation Guide to Media Literacy Education*. New York: Center for Media Literacy.

"Um computador por aluno," http://prouca.comunidades.net/

UNESCO. 2010. Towards Media and Information Literacy Indicators. *Background Document of the Expert Meeting*. Bangkok, Thailand, November 4–6.

Van Dijk, J. 2005. *The Deepening Divide. Inequality in the Information Society*. London: Sage Publications.

Varisco, B. 2002. *Costruttivismo socio-culturale: genesi filosofiche, sviluppi psicopedagogici, applicazioni didattiche*. Roma: Carocci.

Vygotsky, L. 1978. *Mind in society: The Development of Higher Psychological Processes*. Cambridge: Harvard University Press.

Warschauer, M. 2003. *Technology and Social Inclusion: Rethinking the Digital Divide*. Cambridge: MIT Press.

Watts, D. 2003. *Six Degrees*. New York: W. W. Norton.

Chapter 14

Social Media

Transformation of Education

Carolina Duek and Gastón Tourn

Education is one of the most important things for social subjects: it is a preoccupation for parents who want to assure their children the best possible one, it is a governmental concern as the policies targeted both to students and to teachers tend to change and modify the conditions of what happens in the classrooms and, finally, education is a great market in which tools, ideas, and resources are sold as merchandise tending to "guarantee" high standard learning experiences. In sum, when we think about education we are not only considering classrooms, students, and teachers but also families, institutions, sociability, and governmental policies.

In the last few years, a new actor entered the scene: new technologies. The speed with which new electronic devices and media have developed, has changed the communicative eco-system (Martín-Barbero 2003) in which we live. Perhaps one of the most important transformations regarding social bonds is the so-called social media. Not only do they depend on technology but also they rely upon the users that establish different types of interactions using the proposed platforms. Facebook, Twitter, LinkedIn, Orkut, among many others, have transformed the relationship between subjects, technology, and social relationships.

There are some reasons why we claim so. First, the transformation of both time and space (Jameson 1998) and the varied possibilities of being in contact with different people all around the world. This means that with some clicks, we can be in touch with someone from the other side of the planet. We can meet former colleagues, friends, and schoolmates and former boyfriends and girlfriends. Social contact is, in sum, the first transformation. Secondly, the dependency of social media once it was established as the most effective way of presenting ourselves to others (Goffman 1974). Profiles, pictures, thoughts, statements, we can add no matter how many posts to our timeline.

And through the interaction of all the posted items, we configure a certain image of ourselves in front of our "friends." Thirdly, the use of the word "friend" as a synonym of adding someone to our profile by clicking a few times. The idea of friend, as Pahl (2003) maintains, needs to be contextualized as friendship is an informal, dynamic, and variable concept. The type of friends and the relationships we (can) build with them is very related to the material and symbolic conditions in which we live. Friends explain our lives and its changes. We are facing a new use of the word: a friend can be someone we never met but who is an acquaintance of an acquaintance in social networks. The fourth transformation is the one that we are going to analyze in depth: the ways in which new technologies and new media have modified education and the everyday lives of students, parents, institutions, and teachers. New technologies led to new practices and representations of what education should teach students and of what teachers should learn in order to be set to teach in a context that is dynamically changing.

All the afore mentioned leads us to the main issue of this article: what are the transformations that education needs to deal with in a context in which social media aim not only to "conquer" all the available free time from social subjects, but also propose new ways to articulate technology, connection, and education.

We structured this paper around three dimensions and experiences that we want to present and analyze. As scholars from Argentina, we are going to include results from our own research and specific practices that we find in our country so as to illustrate the context in which technology, social media, and governmental policies are being articulated.

"Communication and education" is the first section of the paper and it is there that we will present a theoretical articulation of both concepts in different moments of contemporary media history. Then, "National Program 'Conectar Igualdad'" will present the governmental plan of delivering millions of netbooks to students of public schools in Argentina. In this section, access, democratization, and knowledge are identified and characterized with information about the program. More than 3 million netbooks have been freely distributed all along and around the country. The final section is about "Social media and new platforms." Facebook, Mundo Gaturro (Grupo Clarín S.A. media company), and Club Penguin (Disney International) be analyzed as the most important spaces in and from which children interact with both adults and peers in their everyday lives. The last part of the article discusses conclusions and possible new questions that our research has raised in its course.

The hypothesis that we will sustain is that new technologies have had a major impact in education, and social media have definitely transformed the everyday practices of social subjects. In this context, education should be rethought and reconceived in order to pair its discourses with the new context.

The reception of these new technologies in Argentina has been really unequal and uneven. The impossibility of dealing with the new characteristics of the context can be explained by political, economic, and cultural factors. Those factors are analyzed in the three sections of this paper.

COMMUNICATION AND EDUCATION: A TENSE, DYNAMIC, AND PRODUCTIVE RELATIONSHIP

García Canclini (1995) presented a hypothesis that changed the ways in which citizenship and consumption are thought of in Latin America: he claimed that in a (neo)liberal context, consumption was a way of sending meaningful messages to others. Through our consumption, we constructed our identities. Citizenship turned out to be an old concept that was no longer productive to think about the constitution of identities in a market economy and in a world in which money and status were strictly related. Even though García Canclini's books and researches have been translated and distributed all around the world, it is indeed necessary to question his statements: in an absolutely uneven region such as Latin America, to erect consumption as a way of constructing identities is polemic, to say the least (Rigal 2004). It is also problematic as it claims that people who are beyond the market (as they live in precarious conditions) could not be considered as a part of this new context. But besides our observations (that are mostly ideological and conceptual) there is in fact a new social subject that appeared with a definite strength in the 1990s in Latin America: the consumer, the subject who participates in the global and national market by buying, selling, and accumulating money and goods (Ford 2000).

It is in this context where new technologies of information and communication appeared. Tedesco (2000) claims that in this scenario both knowledge and education need to be rethought. It is in the new consideration of how knowledge is and can be built that we need to rethink the ways in which education can or should change. It is not necessary to transform the whole educational system in the light of new technologies, new ways of connections and new ways of getting to substantial knowledge. As Simone (2001) affirms, in a world in which there seems to be a huge amount of available information, knowledge is always far from being democratic or accessible. The world moves in a direction in which those who have the tools or the possibility of accessing databases that accumulate millions and millions of pieces of information, are the only ones who get to access the relevant knowledge that makes a difference (Piscitelli 2002; Martín-Barbero 2000).

It is in this specific frame that we need to consider the relationship between communication and education: communication and education are the two most important tools to make a difference in the contemporary world. Children who go to school, have any type of family or adult in charge, children who establish relationships with peers and with teachers; those are the ones who can make a difference in the future world. Education needs communication to raise children who will be able to analyze, identify, and criticize media messages and new technologies from a complex point of view.

That should be the objective of education: to help social subjects to build a complex and analytical view of the world they live in with varied tools and instruments. Freire (1986) sustained that by making questions and receiving questions subjects learn how to question their environment. And that is why, as a first provisional conclusion, we want to claim that education does not only happen at schools or special institutions. Education is a group of practices and representations, of histories, culture and traditions, of discourses and folklore (as Gramsci, 1992 presents it). The reflection upon new technologies and education needs to begin by acknowledging this: education is what happens along all our lives in varied and diverse contexts.

Buenfil Burgos (1991) identifies an educational process by an interpellation, something that includes a new content that might modify the everyday life practices in terms of a transformation or a reaffirmation. This gives us the idea that any specific action can be educational if a subject is interpelled by someone else to transform or, in the oppositional form, defend a previous idea. The subject, according to Buenfil Burgos, needs to "accept" the invitation to what it is proposed. And that is how education can be a factor of change and transformation. Questions, new ideas, values, and models of behavior: all these locate education in the global social frame in which each practice can be as meaningful as the involved subjects decide it to be. In sum, education is a social practice, and communication is the major tool from which educational processes occur (see Watzlawick, Beavin, and Jackson 1973).

If we stated and justified that education is beyond classrooms, and communication is the main tool to build educational practices and processes, it is high time we reflect upon new technologies and social media. To be in touch, to be "communicated" seem to be the two most important values for connected societies. Most people who work need to handle electronic devices, computers, and gadgets that guarantee both the efficiency and the control of human actions (that, as we all know, are far from being perfect). The fact that we are permanently controlled by others was analyzed by Foucault (1975) in his brilliant research about surveillance. We act as if we were being watched, we *act*, we display in front of others, series of actions that might help us be who we want to be. New technologies contribute both to the control and to the efficiency of the system.

Ritzer (1997) characterized McDonaldization as the extension on a global scale of the principles that organize McDonald's as one of the most successful businesses in the world to other domains. Perhaps today the main actors that replace the role of Mc Donald's in the 1990s are social networks and social media. All our online actions are of public domain, even though we think our profile is only seen by people authorized by us. Our online life turns out to be customizable and controlled by others. We play, we post thoughts, we present our relationships, and we interact with others through platforms that engineers and media specialists designed for *us* to make *them* rich. We must never forget this: even when we feel we are "free" to look for whatever we want to, there is a system that organizes online searches and the ways in which results are presented.

Control, surveillance, and power are intersected in almost invisible structures. And the possibility of making public all these strategies, mechanisms, and ways through which power acts silently remains on education. Contemporary social subjects *need* to have certain tools and concepts to confront the system that becomes more and more closed around them (Huergo 2000). Employees are permanently controlled by their bosses on their online interactions, Facebook and other social media profiles are constantly checked by superiors; we are raising children in a world in which control is even stricter than the one that Orwell imagined in *1984*.

All the afore mentioned is closely related to communication: in the exact same time in which media say that we live in the era of communication, the mere concept of communication seems to be built only as a synonym of being online and connected. Even though we are perfectly aware that communication is far from being only that, the contemporary hegemonic discourse seems to reproduce this idea. Pierre Bourdieu (2001) maintained that communication was a very complex concept and, to understand it, we had to locate it in a "market" in which meanings and opinions encountered in different power relationships. In this way, the subject who could explain or present the arguments better or make others pay attention had more "capital" in that market. To think about communication as a market is a very productive way of reflecting upon contemporary relationships in a digital environment.

Each action in our everyday lives means something beyond its immediate interpretation. That means that we always act trying to be in a valued and cherished position in front of others. Language is one of the most important dimensions in the market that Bourdieu describes. He claims that the ability of creating and presenting discourses to others locates us in a privileged position in that exchange. This can be perfectly thought when analyzing education, social media, and subjects. Education, we have already said, involves an interpellation and an acceptance by the recipient subject of the fact that something new is going to be received so as to reinforce, change, or argue with previous conceptions. Barthes (2002) says that language is a "skin," and it establishes contacts with others through it. Language is, in conclusion, the key to understand contemporary culture and communications but also (and, perhaps, mainly) the key of social transformation that should be used to improve capacities of social subjects in an electronic and oversaturated environment.

Language as a skin, communication as a synonym of connection, education as a voluntary activity that tends to widen, reinforce, or contradict previous conceptions: a synthesis of this first section of the paper and an introduction to the second one. Constructing an approach to social media and education without considering the importance and relevance of contemporary conceptions of subjectivity and communication would have led us to conceptual misunderstandings. And that is, certainly, not one of our objectives.

PUBLIC POLICIES, NEW TECHNOLOGIES, AND EDUCATION: THE NATIONAL PROGRAM "CONECTAR IGUALDAD"

We have established the main theoretical bonds between communication and education, tracing the concept map that will guide us on the interpretation of the following experiences. Although these cases are focused mainly in Argentina, some patterns can also be found in other latitudes. Moreover, we think that our specific context—a developing country—can allow us to understand distinctive aspects concerning new technologies and education systems.

First, we will analyze the ways in which information and communications technologies (ICT), and more precisely social media, are introduced in Argentinian schools by the national program "Conectar Igualdad" (in English, "Connecting Equality"). This program was carried out by the Argentinian government, and it has already delivered more than 3 millions netbooks to students of public schools in Argentina.[1] We will revise the objectives of this program, assess the role that social media plays in it, and see how schools interacted (or not) with these new platforms.

"Conectar Igualdad" was launched in 2010 with national coverage for public schools (as well as vocational, special education ones, and classrooms in hospitals). It has become the largest 1:1 initiative in the world. The so-called 1:1 model, based on or inspired by the "One Laptop Per Child" proposal by Negroponte (2008), is a socioeducational paradigm which aims at both equity and quality, diminishing the digital and pedagogical divides, and transforming the learning process based on the provision of one computer per student (and teacher), independently of the educational objectives, methodologies, or approaches (Bianchi and Esnaola 2011).

The first country in the world to adopt this model was Uruguay, which placed an order for 100,000 laptops in October 2007 (Uruguay has 3,395,000 inhabitants). Argentina soon caught up and became the largest promoter worldwide. The program is financed by Argentina's presidency and Ministry of Education and is one of the most important public policies that the government has taken during Cristina Fernández de Kirchner's administration. "Conectar Igualdad" is constantly exhibited as one of the merits of Kirchner's political party at different mottos and propaganda.

Oszlak and O'Donnell (1994) define public policies as a set of actions and omissions that show a certain modality of intervention of the state regarding a matter that gathers interest from other actors of civil society. In this sense, we will analyze "Conectar Igualdad" as a public policy both in terms of its actions and its omissions. This dual perspective will allow us to interpret how the Argentinian state understands ICT and the representations that circulate around new technologies in a developing country.

The basis of this program established the following objectives:

1. Promote equal opportunities among children and young people from Argentina, giving them an instrument that reduces the digital divide.
2. Construct a universal policy of digital inclusion with a federal scope.
3. Guarantee that every Argentinian can access the best technological resources and information.
4. Educate responsible subjects, capable of using knowledge as a tool to understand and transform constructively their social, economic, and cultural environment.
5. Develop skills to manage new languages brought out by the ICT revolution. In this sense, give students more opportunities to find a job.
6. Improve teaching and learning processes with ICT by modifying the ways of working in class and at school.
7. Accompany families to make them participate actively in the learning process of their children.
8. Strengthen teachers' training to take advantage of ICT in class.[2]

If we follow these eight objectives we might be led to conclude that, "Conectar Igualdad" is a progressive policy that is helping many children from disadvantaged classes to gain access to new technologies and experience their new languages. However, there are some analytical issues that have to be considered.

First of all, the program does not differentiate quantitative from qualitative access to ICT. The basis of "Conectar Igualdad" equals having a laptop to being enriched by technology. As Bourdieu (1973) has already stated, while one can possess objectified cultural capital by owning a book or a device, one can "consume" the object (take advantage of its cultural implications) only if one has the proper foundation of conceptually and historically prior cultural capital, whose transmission does not accompany the access to that object.

Therefore, even if delivering 3 million netbooks improves and facilitates interaction with ICT, it does not automatically provide an equal significant access to these technologies.

Tiramonti (2004) points out that neoliberal policies and economic crisis have undermined a common referent in Latin American education systems. Even if the "golden age" of capitalism (Hobsbawm 1994) was organized according to segmentations divided by socioeconomic sectors, behind that logic survived a *continuum* in which disadvantaged students could aspire to rise in the social scale. On the contrary, contemporary fragmentation implies the existence of strongly different worlds. In the education system, Tiramonti (2004) notices the presence of divided routes without a common ground. Latin American schools tend to be more varied between each other: there are institutions for elite students, those who are specialized on the arts, those who are focused on foreign languages. However, this heterogeneity is diluted when we see the type of population they embrace, since internally each school tends to include students from similar socioeconomic statuses.

Baudelot and Leclerq (2008) emphasize that quantitative democratization is not a synonym of qualitative democratization. Quite the contrary, inequalities become deeper when school attendance increases. The education system becomes a space for all but not everybody has the chance to gain access to the same quality of education. In this sense, school produces an inclusive exclusion, since all students are included within educational institutions but some follow certain careers that only guide them to social segregation.

In a context of educational fragmentation (Tiramonti 2004), technological equality does not guarantee a pedagogical leap forward. Inequalities relative to different cultural capitals still condition what students can do with a device. "Conectar Igualdad" supposes a big step forward in terms of quantitative access, but it does not pay attention to persistent inequalities concerning different capabilities and cultural backgrounds. Laptops do not reduce by themselves the educational divide.

Which is the role that social media plays in "Conectar Igualdad"? Popular social networks such as Facebook, Twitter, or Google+ are not generally introduced as a teaching resource or a pedagogic tool. On the contrary, most teachers think that they are a distraction that has to be either restricted or ignored. Social media is understood only as a component of leisure time. In some disadvantaged urban areas, children already had access to electronic devices such as mobile phones and TV. For them, owning a personal computer was something new but they previously interacted with social media on their smart phones: "If they don't let me to use Facebook on my laptop, I can always use my mobile phone" (Argentinian student, 11 years).

In this case, teachers did not take advantage of previous practices to go forward and strengthen students' interests through a critical approach. Social networks have been mainly regarded as a waste of time. This lack of pedagogic purposes linked to social media can be perceived in the educational aids that were prepared by the Ministry of Education. The basis of "Conectar Igualdad" emphasizes that this program does not maintain merely an instrumental perspective of ICT: "Education with ICT is not only the instrumental use of new technologies. It also implies learning new competencies relative to information management, communication, interaction with other cultures in a globalized world, innovation and regular updating."[3]

"Conectar Igualdad" claims an integrated approach to new technologies. However, in the official webpage with pedagogic resources for teachers there are only courses to learn how to use an Office software suite. Moreover, it is only taught how to manage Microsoft Office, the commercial set of desktop applications developed by the company founded by Bill Gates. This fact seems to be contradictory with the principles of the governmental program that supposedly defends free software. Stallman (2011) has publicly criticized this issue and renamed "Conectar Igualdad" as "Conectar Maldad" ("Connecting to Evil"). Therefore, the program does not propose any activity to incorporate social media as a tool for significant learning. Social networks are disqualified as a medium of entertainment, when they could be used as a resource for consulting information and learning collaboratively.

How is knowledge conceived in the educational materials designed by "Conectar Igualdad"? The official presentation declares that this program understands knowledge as a collaborative production that allows students to transform constructively their social, economic and cultural environment. But again, there is a gap between what is asserted and what is purposed to teachers. All suggested activities conceive Internet mainly as a tool to look up information, instead of taking advantage of its interactive features. The web does not seem to be a place to construct collaboratively knowledge with other users. According to this pedagogical manual, Internet is only an improved version of a library.

In a few words, delivering netbooks has not yet changed the dynamics of traditional education. The work is not done in networks since most activities still pursue a narrow approach to knowledge. Cabello (2006) states that new technologies collide with a different logic when they are introduced at schools. Education systems are still shaped according to the organizational patterns of taylorism, based on serialization and the disciplinary control of bodies (Foucault 1975).

Boltanski and Chiapello (2002) identify that this new stage of capitalism is led by a post-bureaucratic logic that appeals to a different organizational model. This arrangement comes into conflict with traditional institutions such as schools. Even when schools try to introduce new dynamics, they still readjust them to its disciplinary mold. This can be observed in Edmodo, a social network that was developed for schools and is promoted by "Conectar Igualdad." Edmodo is a social learning platform for teachers, students, and parents and has more than 20,000,000 users worldwide. It is called a "social network," but its web architecture is the opposite from social media. Edmodo reproduces a "verticalist" approach of communication where teachers are the only users that can take decisions over their students' interactions. Teachers have to accept students' comments before they are posted. This authoritarian and unilateral model of communication is the opposite from social networks. In this sense, Edmodo only brings online the organizational culture of traditional education.

In sum, delivering 3 million netbooks increased the use of social media at schools but schools did not take that into account (social networks were either forbidden or disregarded). However, this practice was not ignored by students. Social media in Argentina plays a major role. According to comScore, Argentina ranks first in worldwide social networking engagement at nearly 10 hours per visitor each month.[4] Schools did not notice this huge use of social media, but researchers cannot make the same mistake. In the last section of this article, we will analyze the most popular social platforms in Argentina: Facebook, Mundo Gaturro and Club Penguin.

SOCIAL MEDIA AND NEW PLATFORMS

We claimed that social media has transformed the everyday practices of social subjects. In this final section, we will analyze what children actually do when they interact with new platforms. We will identify which particular types of relationships, actions and dynamics are promoted within social networks.

The Argentinian Ministry of Education conducted a survey in 2006 and not a single kid mentioned that he or she used a social network (Morduchowicz 2012). A new research carried out by the Ministry 6 years later, in 2012, showed that more than 95% of children have visited this type of web sites. Moreover, 75% had a personal profile on Facebook, the most popular social media site in Argentina (Morduchowicz 2012). A typical visit to the platform developed by Mark Zuckerberg lasts around 28 minutes.[5] According to comScore, Latin Americans dedicate to this type of web pages 56% more time than the global average.[6] In this context in which social networks proliferate, an aspect that was not deeply explored by researchers is the role that social games play in the development of these digital platforms. Currently, playing is the fifth most popular activity on Facebook.[7] Moreover, from the top 20 of most used and best qualified apps, 18 are videogames.[8] Even if these games are not spread with the speed that other tools reach in this social network, they achieve an extraordinary loyalty since game users tend to play frequently (Kirkpatrick 2010).

At the same time, the most popular social media sites among Argentinian children, apart from Facebook, are focused almost exclusively on social games. In the Argentinian case, Mundo Gaturro and Club Penguin constitute two trendy web pages to play on the Internet. Mundo Gaturro is a platform based on the characters created by cartoonist Cristian Dzwonik (Nik) and it is financed by CMD (a company of Grupo Clarín S.A., the largest media group in Argentina). Club Penguin was founded in October 2005 and it is a virtual world developed by New Horizon Interactive (an enterprise that was acquired by The Walt Disney Company in 2007). In this social network, some penguins take part in different games in an environment similar to Antarctica. Both web sites prioritize social games over any other digital practice.

Vygotsky ([1930] 1986) claims that games are a lively school for children. Their importance is vital to shape children's personality and world view. Nowadays, games are increasingly migrating to electronic devices. In this context, analyzing social games is decisive to understand this "lively school" for contemporary childhood.

Social games share some specific characteristics that differentiate them from other means of electronic entertainment. O'Neill (2008) provides a definition of this type of activity: "Social games are a structured activity which has contextual rules through which users can engage with one another. Social games must be multiplayer and have one or more of the following features: turn-based, are based on social platforms for providing users with an identity and are casual."

Social games developed for Facebook, Mundo Gaturro and Club Penguin present many common patterns. In all these platforms, gamers find virtual worlds of simulation full of arcades. Social media sites create a digital environment where users can connect to each other. Within those simulated settings, gamers can play to different videogames with simple graphics. Arcades give preference to the simplicity of fast actions and the repetition of interactions.

Sarlo (1994) defines videogames as fictions of "action without story," while they introduce plots where everything happens in terms of movement, but nothing is left relative to the tale. Urresti (2008) adds that this entertainment requires a speed and plain reading, more visual and concrete than discursive or symbolic, more immediate and instant than complex or thorough. Although this characterization largely corresponds to the dynamics present in social games, we should clarify that not all digital games prioritize rapid movement over the narrative component. By contrast, videogames for consoles mostly elaborate fictional frameworks in which, to advance to the next level, one should pay attention both to the agility of execution and to the story. Arcades are the only games identified by the presence of low-resolution graphics and a layout of simple goals. Videogames for social networks have, thus, lower quality in graphic and narrative terms, compared to other industry products. Because of these limitations, they display a "casual" gaming experience. In this sense, they discourage users' immersion into the game's plot since they demand an intermittent and scattered attention.

The final objective of every social game lies in obtaining more objects and a greater scoring than the opponents. The logic of these games is cumulative: one should always possess more digital goods, better furniture and clothes of higher quality. These virtual apps display a teleology of acquisition, in which interactions are meaningful if gamers make some profit. In Mundo Gaturro, for example, most of the announcements within the game insist that one can get items that are more expensive and stylish with a VIP membership. Cantora and Molinari (2012) emphasize that the status on this platform has to do with something objective and material: the number of objects that the user has managed to acquire throughout the game, whether playing for coins or entering a credit card which gives purchasing power that has a basis in real money.

If the aim of the game is to accumulate objects, the dynamics is based on showing what you get. In Farmville,[9] for example, the app suggests that all player acquisitions should be published in the timeline of each user. This exhibitionism is also found within the social game. As soon as you enter the app, a list of scores shows your place in connection with the rest of your "friends." These statistics reach a level of precision that details who has won more points daily, weekly and monthly. It seems that the only possible ontology in these games is exhibition. In social games, to be is to be valued through possessions.

Thus, we notice that social networks primarily display a game mechanics that is both cumulative and exhibitionist. This mechanism of virtual platforms implies a player/child who must purchase and flaunt objects if he or she wants to have fun (Duek et al. 2012). Social networks do not develop recreational environments but spaces of peer competition, organized by the amount of assets held by each participant.

What types of digital relationships foster these websites? Social games seem to recover some cooperative mechanisms. In the "Help" section from the Disney platform, one of the key questions states: "I am new to Club Penguin, how do I make friends?." Under this title, they recommend certain actions that seek to promote mutual links between users:

The best way to make friends is to get involved and be kind to others.

- Waddle up to someone and say hello or smile and see what happens.
- Ask another penguin to play a game of Sled Racing or Find Four.
- Invite other penguins to an igloo party.
- Become a Tour Guide and show new penguins around the island.[10]

These interactions contribute to users' cooperation and mutual collaboration. For instance, when beginners go to the mountain of this virtual island, they can be oriented by advanced players (who act as "tour guides").

According to Mark Zuckerberg, founder of Facebook, the main effect of social networks is that they favor a "gift economy."[11] For the CEO, these platforms promote an exchange of goods and services without an explicit agreement of *quid pro quo*.

However, in social games these reciprocal exchanges are used as a means to promote utilitarian relations between users. All interactions promoted in these games are based on a "self-interested" reciprocity. From sending gifts to other "friends," you get points that allow you to obtain new objects. Thus, while this entertainment recovers certain reciprocal mechanisms, these mutual links respond to a competition against others. Unlike potlatch that is governed under the principle of loss and unproductive expenditure (Bataille 1987), social interactions in these games are guided according to a highly rational and calculated logic. Reciprocal mechanisms are based on a speculative system that serves selfish return. Social games do not promote real cooperation, but they use collaboration as a resource to promote competition among players.

In short, the architecture of social networks strengthens certain patterns linked to consumerism and exhibition. In social games, it is clear that not all players start the game in the same step. In order to fully participate in these videogames, players should purchase a VIP membership or virtual money. The availability of financial resources is what positions users at the top of the rankings of games, better dressed and most productive crops.

Social games have important educational effects since they play a significant role in the subjectivation process of contemporary childhood. In a recent empirical study, Morduchowicz found that 70% of Argentinian children who live in privileged households and 85% of disadvantaged areas use a PC to play (2008, 47). Considering the exponential growth experienced by social networks in the last years, it is possible that most videogames will progressively converge on these websites. Even the consumption of other media overlaps the use of these virtual platforms. According to a research carried out by Ericsson ConsumerLab in late 2012, 62% of TV spectators in the world visit at the same time social networks. This represents an increase of 18% from the simultaneous use of social networks and TV that had been relieved in 2011.[12] If these trends continue their current movement, it is feasible to expect that digital platforms will possess a significant prominence in the communicative eco-system of the following years. This means that social games will be a central cultural consumption that schools will have no choice but to finally notice and work with them.

SOCIAL MEDIA AND EDUCATION: TRANSFORMATIONS IN THE INFORMATION AGE

This article has been divided in three different parts so as to organize our exposition. Communication and education, the National Program "Conectar Igualdad" and finally social media and new platforms, were the topics through which we presented the context in which education is facing major challenges.

We began and we are going to finish by emphasizing that social media already have an impact and an effect on education. That is, even though formal education is reluctant to include and accept the importance of digital media for both students, professors and parents, these new platforms operate on a daily basis in socialization and on the relationship with peers and significant adults. We might say that new technologies, devices and spaces compose a hidden curriculum. And the major problem with hidden curriculum is that, even though we cannot formally "see" it, it constantly transforms communication and education in these new scenarios.

Giroux (1996) defines hidden curriculum as both a possible intention from education to transmit something without explicitly mentioning it or as an element beyond the educational field that impacts in a specific way in the curriculum (Ball 1990). In this case, we would be stating that the second possibility is the one related to social media. In a few words, Facebook, Mundo Gaturro and Club Penguin operate on the social discourses that students take to their classrooms and to their social relationships every single day of their lives. In a different moment of recent history, this role was occupied by television as the main source from which social subjects built their bonds with others and with the context in which they lived (Duek 2013).

Nowadays, in a very short period of time, mass media have lost their space against digital platforms and new technological devices. And that means not only that "traditional" media were defeated but also that new spaces and discourses have taken a hegemonic role in contemporary urban societies. The discourse about digital media and the impact on social status of new devices configure new identities and relationships. For children this transformation has had a double impact: on their everyday lives and in the ways in which they become social subjects and also in their relationship with their teachers and significant adults. The transformation of knowledge and the abilities regarding ICT (generally attributed to students) have displaced the authority of teachers and professors. We are not saying that adults have lost their power nor that the vertical relationship in the educational process has been modified. We are only mentioning that the access to both significant and irrelevant knowledge is now closer to students than ever. The truth is that media constantly help to crystallize this new way of representing the relationship between students, teachers and parents. It is highly convenient to construct children as independent and autonomous in front of new media to reproduce the system as it is. That means that if the market identifies children as a target, they can sell them specific products, tools and apps.

We would like to conclude by highlighting the responsibility of national governments regarding access to technology and knowledge. "Conectar Igualdad" has been a major plan with a great impact in the everyday lives of both teachers and students (and to their families as, in many cases, the netbook from this program was the first and only one in the household). However, this plan has not conceived a critical approach (or any approach at all) to the possible uses of ICT and their pragmatics. In other words, access has been mistaken as the only right while students, who already had a computer at home, had a clear advantage in the use they could give to the device. We are demanding a qualitative democratization that *has to* include a complete and articulated training in varied platforms and digital languages. The co-construction of parents, teachers and students into critical readers and users has to be the challenge of the near future. If this is not achieved, the delivery of millions of netbooks will fall in a void, a non-critical void that will only reproduce previous social inequalities. To keep away from this, we should defend a democratization not only of devices but also of knowledge. That is, without any doubt, what public policies should guarantee in this Information Age: access, knowledge and equality.

NOTES

1. The updated number of delivered netbooks can be checked at the official webpage of "Conectar Igualdad": http://www.conectarigualdad.gob.ar/seccion/sobre-programa-6
2. The basis of this program can also be checked at the official webpage of "Conectar Igualdad": http://www.conectarigualdad.gob.ar/seccion/sobre-programa-6
3. Also available at the official webpage of "Conectar Igualdad": http://www.conectarigualdad.gob.ar/ seccion/sobre-programa-6
4. This statistics belongs to a study carried out in 2012 by comScore: http://www.comscore.com/Insights/ Press_Releases/2012/12/Argentina_Ranks_First_in_Worldwide_Desktop_Social_Networking_Engagement
5. This data was gathered by Alexa. The updated metrics for Facebook can be seen at http://www.alexa.com/ siteinfo/facebook.com
6. This statistics comes from the study carried out by comScore in 2012.
7. *eMarketer* declares that almost half of Facebook users employ this social network for playing. In this sense, it is the fifth most popular activity, even above watching videos (cf. http://www.emarketer.com/ Article.aspx?R=1007665).
8. The top 20 of Facebook apps is available on https://www.facebook.com/appcenter

9. Farmville is a social game developed by Zynga and has become the most popular app in Facebook with 81,844,829 monthly active users. The goal is to manage a virtual farm: planting and harvesting crops and raising animals. The user needs to obtain electronic money (Farm Cash) to buy objects and, thus, increase his or her level.

10. These recommendations can be found at the Help Center: http://www.clubpenguin. com/help/help-topics/friends/im-new-club-penguin-how-do-i-make-friends

11. Mauss states that the gift economy is a total benefit system where loans "are done voluntarily, through presents or gifts, although in essence they are rigorously mandatory" ([1924] 2009, 75). The French sociologist analyzes in particular reciprocal exchanges practiced by indigenous people in northwestern North America, which take the form of a ceremony called potlatch.

12. This data corresponds to a research carried out by Ericsson ConsumerLab in December 2012: http://www.ericsson.com/res/docs/2012/consumerlab/10-hot-consumer-trends-2013.pdf

REFERENCES

Ball, S. 1990. *Politics and Policy Making in Education: Explorations in Policy Sociology, Vol. 1*. London: Routledge.

Barthes, R. 2002. *Fragmentos de un discurso amoroso*. Buenos Aires: Siglo XXI.

Bataille, G. 1987. "La noción de gasto" in *La parte maldita*. Barcelona: Icaria.

Baudelot, C. and F. Leclerq. 2008. *Los efectos de la educación*. Buenos Aires: Editorial del Estante.

Bianchi, G. and G. Esnaola. 2011. *Formación académica en Tecnología Educativa y el Modelo 1 a 1 en Argentina*. Sevilla: XIX Jornadas de Tecnología Educativa.

Boltanski, L. and E. Chiapello. 2002. *El nuevo espíritu del capitalismo*. Madrid: Akal.

Bourdieu, P. 1973. "Cultural Reproduction and Social Reproduction" in *Knowledge, Education, and Social Change: Papers in the Sociology of Education*. London: Tavistock.

Bourdieu, P. 2001. *Poder, derecho y clases sociales*. Bilbao: Desclée.

Buenfil Burgos, R. 1991. "Análisis del discurso y educación" in DIE-CINVESTAV-IPN, September 1991.

Cabello, R. 2006. *Yo con la computadora no tengo nada que ver. Un estudio sobre la relación de los docentes con las tecnologías informáticas*. Buenos Aires: Prometeo-UNGS.

Cantora, A. and L. Molinari. 2012. *Jugar a "Mundo Gaturro." Acerca del juego, la exhibición y el consumo en la plataforma online*, thesis in Communication (University of Buenos Aires). Buenos Aires: FCS-UBA.

Duek, C. 2013. *Infancias entre pantallas. Las tecnologías y los chicos*. Buenos Aires: Capital Intelectual.

Duek, C., N. Enriz, F. Muñoz Larreta, and G. Tourn. 2012. "Juego, redes sociales e infancia: hacia la definición de nuevos escenarios comunicativos" in *V World Congress for the Rights of Children and Adolescents*, San Juan.

Ford, A. 2000. "Procesados por otros. Diferencias infocumunicacionales y sociocultura contemporánea" in *Comunicación-Educación. Coordenadas, abordajes y travesías*. Bogotá: Siglo del hombre Editores.

Foucault, M. 1975. *Vigilar y castigar. Nacimiento de la prisión*. Buenos Aires: Siglo XXI.

Freire, P. 1986. *Hacia una pedagogía de la pregunta. Conversaciones con Antonio Fernández*. Buenos Aires: Ediciones La Aurora.

García Canclini, N. 1995. *Consumidores y ciudadanos. Conflictos multiculturales de la globalización*. México: Grijalbo.

Giroux, H. 1996. *Placeres inquietantes*. Barcelona: Paidós.

Goffman, E. 1974. *Frame Analysis*. Boston: Northeastern University Press.

Gramsci, A. 1992. *Antología*. México: Siglo XXI.

Hobsbawm, E. 1994. "The Golden Age," in *The Age of Extremes: A History of the World, 1914–1991*. New York: Pantheon Books.

Huergo, J. 2000. "Comunicación/Educación: itinerarios transversales" in *Comunicación-Educación. Coordenadas, abordajes y travesías*. Bogotá: Siglo del hombre Editores.

Jameson, F. 1998. "Preface" in *The Cultures of Globalization*, Jameson and Mayoshi (eds.), Duke University Press.

Kirkpatrick, D. 2010. *The Facebook Effect*. New York: Simon & Schuster.

Martín-Barbero, J. 2000. "Ensanchando territorios en Comunicación/Educación" in *Comunicación-Educación. Coordenadas, abordajes y travesías*. Bogotá: Siglo del hombre Editores.

Martín-Barbero, J. 2003. "Retos culturales de la comunicación a la educación. Elementos para una reflexión que está por comenzar" in *Comunicación, medios y educación*, Roxana Morduchowicz (comp.). Barcelona: Octaedro.

Mauss, M. [1924] 2009. *Ensayo sobre el don. Forma y función del intercambio en las sociedades arcaicas*. Buenos Aires: Katz.

Morduchowicz, R. 2008. *La generación multimedia: significados, consumos y prácticas culturales de los jóvenes*. Buenos Aires: Paidós.

Morduchowicz, R. 2012. *Los adolescentes y las redes sociales*. Buenos Aires: FCE.

Negroponte, N. 2008. "One laptop per child," in *American Association for the Advancement of Science* (Lecture). Retrieved 1/4/2008.

O'Neill, N. 2008. "What Exactly are Social Games?" http://www.socialtimes.com/2008/07/social-games

Oszlak, O. and G. O'Donnell. 1994. "Estado y políticas estatales en América Latina: hacia una estrategia de investigación" in *Para investigar la Administración Pública*, Kliksberg and Subirandt (comps.). Madrid: INAP.

Pahl, R. 2003. *Sobre la amistad*. Madrid: Siglo XXI.

Piscitelli, A. 2002. *Ciberculturas 2.0. En la era de las máquinas inteligentes*. Buenos Aires: Paidós.

Rigal, L. 2004. *El sentido de educar. Crítica a los procesos de transformación educativa en Argentina dentro del marco Latinoamericano*. Buenos Aires: Miño y Davila.

Ritzer, G. 1999. *La Mcdonalización de la sociedad*. Buenos Aires: Ariel.

Sarlo, B. 1994. *Escenas de la vida posmoderna*. Buenos Aires: Ariel.

Simone, R. 2001. *La tercera fase. Formas de saber qué estamos perdiendo*. Madrid: Taurus.

Stallman, R. 2011. "Entre el software libre y la provocación," in *Página/12*. Accessed June 19 2015. http://www.pagina12.com.ar/diario/suplementos/espectaculos/17-22875-2011-09-09.html

Tedesco, J. C. 2000. *Educar en la sociedad del conocimiento*. Buenos Aires: FCE.

Tiramonti, G. 2004. *La trama de la desigualdad educativa*. Buenos Aires: Manantial.

Urresti, M. 2008. *Ciberculturas juveniles*. Buenos Aires: La Crujía.

Vygotsky, L. S. [1930] 1986. *La imaginación y el arte en la infancia*. Madrid: Akal.

Watzlawick, P., J. Helmick Beavin, and D. Jackson. 1973. "Algunos axiomas exploratorios de la comunicación humana," in *Teoría de la comunicación humana*. Buenos Aires: Tiempo Contemporáneo.

Index

About the Editor and Contributors

Barış Çoban is an Associate Professor in the Communication Sciences Department, Doğuş University, Turkey. He call himself an activist-academic. His research interests include social movements, alternative-radical media and political communication. The books he edited, co-edited and translated are: "Discourse and Ideology," "History, Utopia and Rebellion: Sheikh Bedreddin," "Panopticon: Power of Eye," "Media, Nationalism, Violence," "Globalisation, Opposition, Utopia: New Social Movements," "Media, Peace and War," "Reclaim Your Media: Alternative Media," "Alternative Media in Turkey." He has also published in journals in areas of media, sociology and politics. He has lectured on theories of alternative media, social media and political communication.

Chiara Livia Bernardi is a PhD Candidate in the Department of Sociology, University of Warwick. She has vast experience in the world of media and communication, having worked with different clients and across different sectors (from B2B to B2C). She has extensive knowledge of the third sector having worked with universities, NGOs and charities nationally and Internationally. She supported teams of experts in UN projects and had the pleasure to be involved in a project funded by the International Development Research Centre. She has lectured in the UK and Internationally (Turkey, Italy and The Netherlands) and published extensively on Socio-digital cultures, Digital Media and Violence, Hacking and feminism.

Tommaso Gravante PhD in Political Science. Postdoctoral research at UNAM (Mexico) and member of the group of research at Interdisciplinary Group of Studies in Communication, Politics and Social Change, University of Seville (Spain). His research interests revolve around alternative media, emotions and protest, self-organized grassroots movements and socio-cultural change, Latin American study.

Alice Poma PhD in Social Science. Postdoctoral Fellow at UNAM (Mexico) and member of RENISCE (Mexican Research Net on social-cultural study of emotions). Her research interests revolve around emotions and protest, self-organized grassroots movements and socio-cultural change, Latin American study.

Kudus Adebayo is a doctoral candidate in the Department of Sociology, University of Ibadan, Ibadan, Nigeria. His research interests include Nigerian diaspora, institutional vulnerability, urban sociology, and new media and social movement. He has published in edited volumes and peer-reviewed journals.

Abderrahim Chalfaouat is a PhD candidate at the department of Advertising and Communication, Hassan II University of Casablanca. He works on the intersections between television, cultural policy and democratisation in Morocco. He got an MA in Moroccan American Studies in 2011. His research interests include media and society, cultural policy, MENA politics, democratisation and human rights. Abderrahim is an independent writer and journalist with studies and opinion articles featuring at a number of sites in English and Arabic.

Dana Florentina Nicolae B.A. in Art History at the Faculty of History at the University of Bucharest, Romania. Currently getting the Master's degree in Art History and the Philosophy of Culture at the Faculty of History at the University of Bucharest, Romania.

Ilona Biernacka-Ligięza is a Professor of Humanities and she works at the Faculty of Political Studies, Maria Curie-Skłodowska University and Department of Political Studies Opole University. From September 2008 until October 2011, she has been the research fellow of the Institute of Media and Communication, University of Oslo. Her research interests include: local media; mass communication; new media; cross-cultural communication; journalism studies; local democracy; globalization; glocalization; media and politics; media and identity.

Alan Steinberg is currently an Associate Director with the Center for Civic Leadership at Rice University. He earned his Ph.D. from the University of Houston and recently completed a Postdoctoral Fellowship at Sam Houston State University. Alan's focus is on applied political science research and education; including the impact of social media on political engagement and understanding factors influencing space policy.

Simona Stano is Post-Doctoral Research Fellow at the University of Turin and Senior Researcher at the International Semiotics Institute. She holds a PhD in Sciences of Language and Communication from the University of Turin (UNITO, Italy) and a PhD in Communication Sciences from the University of Lugano (USI, Switzerland). During her doctoral studies, she worked as Visiting Research Scholar at the Dept. of Anthropology of the University of Toronto (UofT, Canada), also obtaining a Research Fellowship from the Dept. of Religious Studies. She holds a BA in Communication Studies (UNITO, with 9-month SOCRATES/Erasmus Programme at UV, Valencia, Spain), a MA in Multimedia and Mass Communications (UNITO), and a Postgraduate degree in Intercultural Studies and Peace Culture (UAB, Barcelona, Spain).

Ana Viñals Blanco has a degree in Advertising and Public Relations from the Faculty of Social Sciences and Communication University of the Basque Country, Official Master's Degree in Organization of Congresses, Events and Exhibitions and, nowadyas she is PhD researcher at the Institute of Leisure Studies at the University of Deusto (Basque Country – Spain). She is currently doing her doctoral thesis provisionally entitled: "The e-leisure of youth (16–18 years) of Biscay: identification, influences and educational guidelines" (Granted by the Basque Government).

Her research areas of interest are young people and digital leisure, leisure education, digital literacy, ICT and education, youth leisure policies and digital culture.

Shahriar Kabir is a lecturer of English in Southeast University, Bangladesh. Currently, he is on study leave and pursuing an MA in English literature in USA. In addition to the traditional course of perusals as demanded by his discipline, Mr. Kabir is also interested in reading current cultural events in the light of literary theories. His recent MA thesis is on Sinclair Lewis's Main Street, and he is working on the articulation of voice of the protagonist in the novel.

Anna Rita Popoli holds the position of Public Prosecutor since 2002 with specialization in racial discrimination crimes using the Internet, Defamation on the Internet, On-line fraud and Identity abuse. She obtained her Master of Arts in European Consumer Affairs in 2011: the final dissertation was "Cookies and Electronic Commerce: a Survey About Actual Knowledge of the Issues Concerning Privacy." From September 2014 she is Research Fellow in Internet Law and Social Network at Bologna University (Italy).

Magda Pischetola is Assistant Professor in Digital Media at the Department of Education at the Pontifícia Universidade Católica of Rio de Janeiro, Brazil. Since 2014 she also coordinates the Teachers' training and ICT Research Group ForTec (Formação docente e Tecnologias). She has a Ph.D. in Education at the Università Cattolica of Milan (Italy) and has completed a Post-doctorate fellowship at the Universidade Federal de Santa Catarina (Brazil) in 2012.

Her current research interests are the role of social media in political activism; the impact of technology on literacy practices; the relationship of ICT to social development; digital media and social inequalities.

Carolina Duek has a PhD in Social Sciences from the University of Buenos Aires, a Masters in Communication and Culture. She is a professor at the University of Buenos Aires and teaches theories and practices of communication. She works as a researcher at CONICET where she directs different projects related to children, media and play in the contemporary urban context. She has published three books and several articles in national and international journals. Juegos, uguetes y nuevas tecnologías" (Capital Intelectual, 2014) is her latest book. From 2010 to 2012 she was Associate Professor at New York University.

Gastón Tourn is a Graduate Student from the University of Buenos Aires, currently working for Google London. He had a scholarship with which he wrote his graduate thesis on social media and children. He has published papers in both local and international journals.